AF241744

BURLINGTON AIRPORT

Burlington Airport

GRACE H. PUGH

Copyright © 2026 by Lucy McCullough

All rights reserved. No part of this publication may be reproduced, distributed, or transmitted in any form or by any means, including photocopying, recording, or other electronic or mechanical methods, without the prior written permission of the publisher, except in the case of brief quotations embodied in critical reviews and certain other noncommercial uses permitted by copyright law.

Onion River Press
89 Church Street
Burlington, VT 05401
info@onionriverpress.com
www.onionriverpress.com
ISBN: 978-1-966607-38-0
Library of Congress Control Number: 2026911137

CONTENTS

PREFACE

Grace Pugh was not your typical friend, mother, or elementary school teacher; she was a modern woman with a vision and a passion. She was a pilot, flight instructor, and champion of aeronautics. She was Vermont's First Lady of Aviation and the first female licensed pilot in the state.

Grace was an elementary school teacher in Vergennes in 1931 when she got her first taste of flying at age twenty-three. In early winter, 1932, her husband to be, a friend of her stepfather, landed in the meadow in the front of the family farmhouse in North Ferrisburg. Two years later, on November 10, 1934, she married that handsome daredevil of the sky, Harold Pugh. Harold and Grace were visionaries, and together they dedicated the next thirteen years to the growth and development of the Burlington Airport and the promotion of aviation, a thing they were truly passionate about.

In 1977, Grace was inspired to write *Burlington Airport: The First 25 Years*. In the interest of preserving the memories of this period of early aviation history, her daughters, Patricia Pugh Soutiere and Lucille Pugh McCullough, have agreed to publish the story written by their mother.

INTRODUCTION

I wish I had the ability to express how fortunate I have been to live during those decades that spanned the best of the old and some of the new.

I was privileged to know the coziness of a feather bed. I toasted by the Round Oak heater in the living room with the full hod of coal waiting beside it. Not so enjoyable was the pitcher pump on Sunday night as we filled the huge tubs for the Monday laundry. I remember washing the many, many parts of the cream separator and milking machine: the butter churns (we had both barrel and swing), the outdoor plumbing, haying with horses in the summer, and sleigh rides in winter.

My teaching school days began in a rural school with the common dipper in an open pail of drinking water. To keep from freezing we took turns standing around the wood burning stove. Nearly fifty years later, I was teaching in the most modern of schools with open classrooms, tape recorders, individualization, and behavioral objectives.

When I was a little girl, my father had a 1911 Model T crank-up Ford with a put-down top, brass stay rods and doorknobs, and kerosene lamps. I drove a roadster, with side curtains and rumble-seat, over dusty, washboard roads. People were puzzled and confused when the first traffic light appeared in town. Now as I sit patiently waiting for a chance to blast from my home driveway out onto Williston Road, I think of the speed, the beauty and luxury, the high prices, the parking problems and the traffic. It has been a great age for the automobile.

The greatest change of all has been in aviation. Everything so modern and exciting then seems very primitive or archaic now. Back in the twenties, the pilots were nearly as free as the birds. They landed where they dared, took off where they could, flew as high as they pleased or as low. There were only a few rules. Only the pilots knew what they were, and they didn't always abide by them. They flew by common sense and by the seat-of-their pants. They had few instruments, no radio, no tail wheel or brakes, and often used automobile gas. They had to be "good" to survive. Recognized by their helmet and goggles, they appeared to be the daredevils of the sky, subjects of hero worship.

My first airplane ride was in November 1931, under the watchful eye of my stepfather, Arthur Ashley, manager of the fledgling Burlington Municipal Airport. Ashley had asked a visiting pilot, Walton Waite, to take me up for a look around the city. Flying nearly 500 feet above Burlington and glancing down from the passenger seat of the roaring 220 hp Stinson Tri-motor, I spotted the distinctive pattern of a football field that marked the site of Burlington's Centennial Field. Far below, a swarm of Lilliputian ball players in yellow and dark blue uniforms reminded me of pesky potato bugs on the family farm in North Ferrisburg. The flight was safe and uneventful. I was actually bored. The experience was no different than taking a boat ride.

On a frigid afternoon early winter in 1932, pilot Harold Pugh flew his small plane from Plattsburgh, N.Y. for his annual pilot's physical. Burlington had several physicians authorized to give physicals to area aviators.

Harold's visit to the doctor took longer than he expected. At dusk he left the doctor's office. There was still daylight enough to fly back to New York, but a stiff cold wind whipped across Lake Champlain. Harold attempted to start the plane's engine, but it would not turn over. It had turned so cold he had to put a blow torch under the engine to heat up the oil as that was the only way to start old planes on a cold day.

Harold began to worry that, by the time the engine heated up, he would be making the return flight in the darkness. In those days before radar, it was difficult to navigate in the dark, making flying at night a dangerous proposition. My stepfather persuaded Harold that it was safer to come to the family farm, spend the night, and get a fresh start in the morning. Harold agreed to that, and that is how we met.

That evening, as we sat around the kitchen table chatting, Harold teased me about teaching and my claim of understanding the science of physics. As a challenge, he asked me, "If you were flying from Tennessee to Illinois, and there was a low-pressure area nearby, which way would you fly? Would you fly to the east of the low-pressure area or to the west of it?" With my index finger, I drew an imaginary picture of a low-pressure cell on the table-cloth. I said to him, "Well, you would have to fly to the east of the low-pressure area so you would be able to take advantage of the tail wind." Harold was apparently impressed. A few months later, Harold was working for my stepfather at the airport. One Saturday afternoon, Harold flew an Aeromarine Klemm, an unusual ply-wood monoplane, to the family home in North Ferrisburg, landing in the nearby meadow. He walked up to the house with a big bundle in his arms. It was an oversized flight suit meant for me. He zipped me into the darn thing, and off we went. We circled around Vergennes several times, where I taught school. He brought the plane down over the railroad tracks near our farm, made a tight S-turn, and landed up abruptly on the meadow. Flying in the Aeromarine was the best thing that ever happened to me. It was fun, it was exhilarating, and that is when I knew I wanted to fly planes.

Two years later, Harold and I were married. Flying was something I didn't know much about. I took flying lessons from Curley Mullaney in 1938, and it all became second nature to me.

Grace with flight instructor Curley Mullaney and Jo-Jo, the airport dog

From 1932 to 1945, I was totally involved with the Burlington Airport. It was our home, our work, our hobby, our social life, our only interest, and, for ten years, our only child. We were totally and unselfishly dedicated to the growth and promotion of aviation in all its forms. Like born-again Christians, we wanted everyone to believe as deeply and thoroughly as we did in the future of aviation. We pleaded with anyone who would listen, "Come and watch; take a ride; learn to fly. It's great; it's safe; it's the way of the future." We courted their confidence in every way possible. Our desire for business had to be secondary as there was so little of it. If we were interested only in making a living, our time there would have been extremely short. We wanted everyone to love flying and believe in it as we did.

I would like to recognize the unsung heroes who contributed so much to this budding new science and art called aviation. Only through their interest and support could it have changed so rapidly.

In a few pages, it would be impossible to describe this most exciting and interesting thirteen years of my life. Neither can I adequately point out the rapid progress and many changes that caused that period to be called the Golden Age of Aviation.

THE GOLDEN AGE OF AVIATION

This exciting and miraculous time in aviation history of the Burlington Airport began with the purchase of a metal hangar in 1929, sold by Vermont Airways of Newport, Vermont, represented by Lt. Lyle C. Churchill. Churchill was the first fixed base operator.

In 1931, Stuart Astles was at the airport, assisted by Everett Marden, who later operated the field along Route 7 in Middlebury, with a Stinson bought from Vermont Airways. Arthur Ashley and George Hall came to the field on October 3, 1931. Ashley had a Hesso WACO (a WACO with a Hispanio-Suiza motor). Frank E. Buswell, with his mechanics license and pilot's license, T12020, was at the field for a while also.

Harold Pugh, a resident of Keeseville, New York, had his early instruction from Henry Duffany. Duffany was killed May 21, 1931 when his plane crashed into a house in Plattsburgh. Harold also took instruction from Lt. Lyle Churchill at the Mobodo Airport in Plattsburgh, New York owned by Wilfred Crete. Harold had held a limited commercial license since November 6, 1931. He got his transport license in Burlington in March 1932 and started as the Burlington Airport manager and instructor. His salary was $50 per month, office space when available, and as much hangar space as he wished, unless a cash customer needed it. He kept a field truck and paid a man, or men, to clean the hangars, sweep the ramp, and police the grounds. This did not change from 1932 to 1942 when he began to receive $100 per month and $50 for an assistant manager.

During the winter of 1933, Harold went with Dick Colburn in Dick's WACO F to do instructing at the University of Alabama. In his absence of four months, Jim Beckwith was acting airport manager.

By 1934, the airport was really starting to grow and Harold was all over the place, by car, by truck, or on foot. He was on top of everything: the new administration building and the subsequent enlarging of it, the recurring paving and the lengthening of runways, the building of the wooden hangar, the installation of teletype communications, the radio beam, the lighting of the field, obstruction lights, markers, wind direction indicator, the growing of soy beans to keep the soil from blowing, the courting of Boston-Maine Central Vermont Airways in 1934, and Canadian Colonial in 1939. All this and a rapidly growing flying school! We had 140 students in 1942. Seven-hundred hours of student instruction in July alone. Beginning in April 1943, there were 200 new students per month getting approximately ten hours each. Harold was everybody's hero in those days. Everybody knew him. While Harold was manager of the airport, I wasn't actually a paid employee of the airport, but I did everything I could to help him. I was teaching school at the time, and that kept me pretty busy. I was also taking care of the office and keeping airport records.

The flights averaged one per day in 1931 to a high of 793 per day in 1943. From a grassy, bumpy field, it became one of the best airports in the Northeast and the busiest in the country.

Harold, due to business pressures and politics, resigned as airport manager on June 19, 1943.

On December 30, 1944, the flight school moved all operations to Swanton, where we had also been operating since May with Arthur Viens in charge. We had trained all the pilots the govern-

Harold Pugh

ment needed and all those eligible were in service. Our great success had brought an end to the thing we loved. I was staying at
home to take care of my baby, and Harold was busy fixing up the

family farm in North Ferrisburg into a sportsman's paradise. Ed Steen and Irving Hatin did their very best to keep the airport together but there was no business. The hangar rent and share of snowplowing was $466 per month, which was just fine when we were busy, but at that time it was impossible.

November 10, 1944, Dan Hufnail became Burlington Airport Manager at $200 per month and hangar rent cost him $400 per month. He was a likeable guy, and he did a good job. By then the servicemen were coming back, and many were learning to fly on the GI Bill. I did some ground school for him and some tutoring but soon went back to teaching grade school in Colchester.

In 1946, the postwar surge was in the hands of Hugh L. Finnegan who helped put the airport on a money-making track. He was a political appointee who worked hard and was willing to learn. As alderman, he had been a steadfast supporter of the airport since 1934. Slim Hedman was also of great help.

J. Edward Moran succeeded Hugh Finnegan. This was a political appointment. Moran had just finished a term as mayor. He didn't drive a car—he rode the bus. The airport continued to grow.

The people of Burlington have always been very "air-minded" and supportive of airport projects. This was due to the caliber of the commissioners, the time they spent, their interest, and their sincerity.

The commissioners we knew best were John Burns, Hobart Spear, Alfred Heininger, Ray Crane, Willett Foster, and Fritz Shepardson. They were all hard-working go-getters who would fight (and often did) for anything that would give Burlington a bigger and better airport. Few people knew, and nobody could recount, the numerous ways they contributed to its rapid growth.

AIRPORT COMMISSIONERS

John Burns, World War I flyer
A.D. Butterfield
Hobart Spear, World War I flyer
Ben Lane
Alfred Heininger
Ray Crane
F. W. Shepardson
Willett Foster
I. Munn Boardman
Robert Bromley
Hugh Finnegan
Lester Greene
Dr. Erald Foster
L. J. Walsh
Dick Spear
Roger Fay
Edwin Blakeley
Vincent D'acuti

Newspaper reporters also were among the unsung heroes in our airport development. Everything about flying was news in those days. Anxious as they were for headlines, the reporters usually managed to accent the positive and leave the airport and its personnel as untarnished as possible. They were pilots at heart, and their optimism did a great deal to encourage the interest of the public in the progress of flying. The tone of their remarks was always kind—never harsh or ridiculing. They inspired a just and rightful pride in their airfield.

George C. Stanley, city engineer, was an outstanding contributor to the growth of the airport. It would have been so easy for some people to turn aside, to avoid getting involved, to pretend it wasn't necessary. It would have saved a lot of work and trouble. Not George Stanley! From the very beginning, he was interested. The street department men and equipment did the work. He made thousands of trips to the airport and put in a lot of time there. The runways, the hangars, and all the plans involved a tremendous amount of work and expert know-how. He will never receive sufficient credit for all he did.

My brother, George Lyman Hall, was introduced to aviation by our stepfather, Arthur Ashley. My mother had bought Arthur his first plane in 1931, and he became manager of the Burlington Airport. George stepped into the role of airplane mechanic at the age of fifteen and obtained his mechanics license at the age of seventeen. At age eighteen, he was one of the youngest licensed aircraft and aircraft-engine mechanics in the country. Not only did he work on the small Cub-type planes but was also the local mechanic for Northeast, Colonial, Eastern and Mohawk Airlines. On April 4, 1946, he became Designated Aircraft Maintenance Inspector Number 12. He could go to any airport in the country and "ground" either military or civil aircraft.

The greatest praise of all should go to George Hall and Loren Avery and their men who ran the maintenance department. The fine safety record enjoyed by local pilots and the flying school was due to their skill, thoroughness, and hard work. With planes often flying eight to fourteen hours per day, it wasn't easy to keep each one in perfect condition. The planes were used in all kinds of weather and by many kinds of pilots, but for George and Loren, there was nothing overlooked—no mistakes. All this did a great deal to promote public confidence in aviation, which in turn encouraged the growth of the airport.

George had a sense of humor. The airport lounge was very small and congested at times. Often, loungers would stay too long. George, with a mischievous twinkle in his eye, fashioned a system with a trigger hidden behind the counter to give the lounger an electrical buzz just enough to get them to move along.

George was dedicated to the airport and continued to work there for about forty years. He then worked for the Department of Defense at General Electric until retirement.

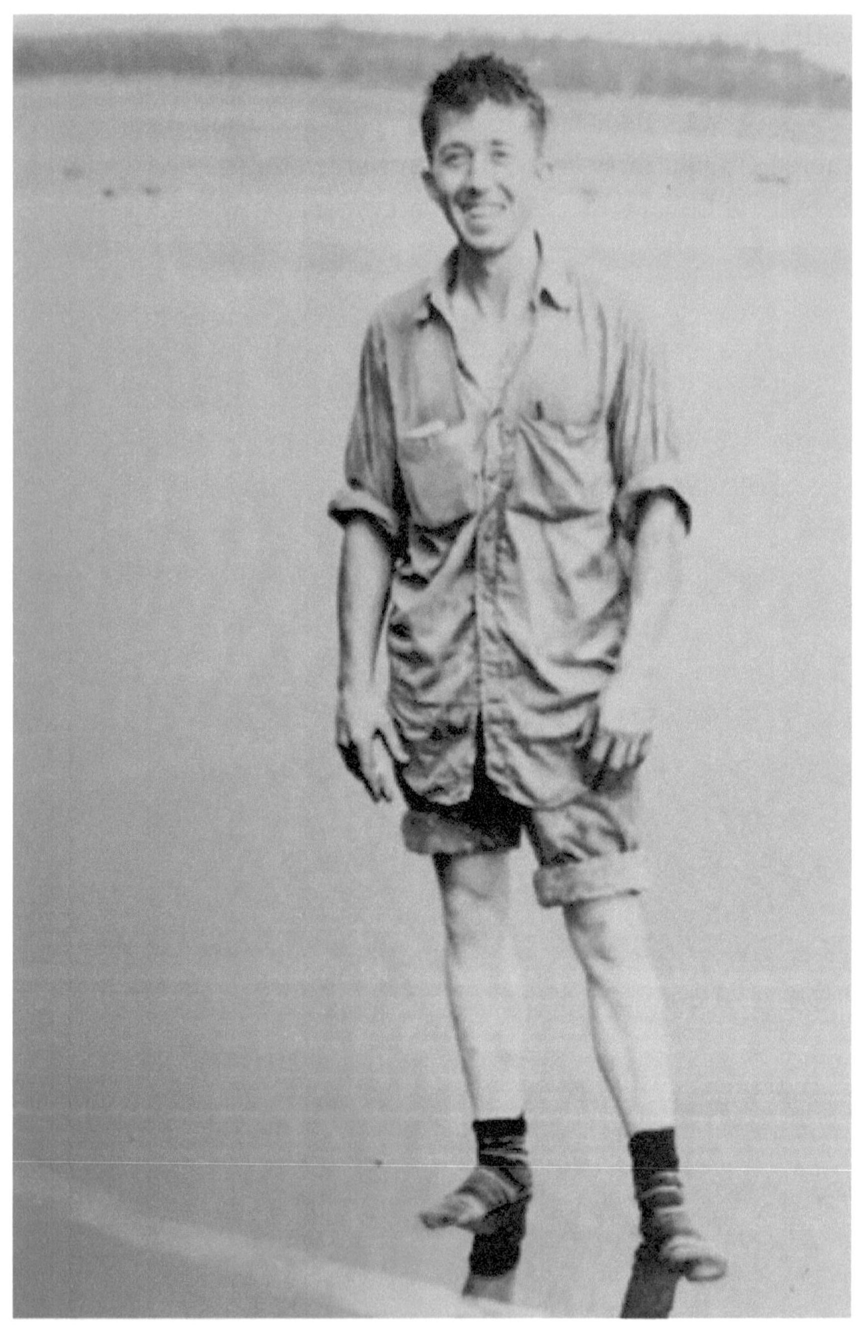

George Hall

We would have folded even sooner than those who preceded us had it not been for my mother, Alice Kimball Hall Ashley. She lived at Bayview Farm in North Ferrisburg. The farmhouse was large with many beds. There were vegetables and canned goods in the cellar, a woodshed full of wood, cows in the barn, and a garden in summer. I can't begin to tell the number of people who called that place home. Those were the dark days of the depression. The rich, the poor, the sad, the sick—they were all welcome. They paid if they could and helped a little. I remember coming home from school and counting twenty-one men's shirts hanging here and there, all freshly ironed. (No perma-press in those days.) She believed in the future of aviation, too, and had been up flying several times before I had been up even once. If she thought she was helping, no personal sacrifice was too great.

Joe and Cora Rock ran the Bristol Airport and were our very good friends and neighbors in every sense of the word. Joe had been a very young Navy pilot in World War I and then became part of a flying circus that toured the country. Cora had taught school in White River Junction for several years. Beginning in 1939, they had a Navy CPT program with students from Middlebury College. Harold and Joe tested each other's students, and borrowed each other's planes, instructors, and spare parts. They were always ready to help when we needed them. Joe was killed landing at Westport after dark, following a heart attack, I'm sure. Joe and Harold both died in 1960.

Some of my good friends and airport companions were those patient, hardworking gals who ran the kitchen-lunchroom: Ina Cushman, Irene Partlow, Zoe (Betty) Blow, Violene Woods. There were others, like Mary Ann Pratt, but the first four stayed about a year and a half each.

In June of 1940, Merle and May Wood came to run the kitchen. They did a really great job. There were lots of people around: passengers, students, and lunches for all. The food was excellent and pleasing to look at. How they served so many so well in such a small area, I'll never know. Merle had just left the farm in Georgia, VT, and was learning to fly. In 1942, he went to Montpelier and became the flying and ground school instructor there. After that, he took more training with Northeast Airlines and became a pilot for them.

Merle Wood played the drum in a band called "Moonlight Serenaders." He was often seen standing with his elbow resting on the side of the refrigerator, sound asleep. Merle became involved with the Country Store at Essex Center and Winooski. He was a fast talker on his one-minute TV commercials.

PEOPLE, PLACES, AND EVENTS

Fifty years-ago today, on the North Carolina sand dune, two bicycle mechanics from Dayton gave the world the airplane.

The trembling contraption of muslin, wood, and wire that Wilbur and Orville Wright crafted of whole cloth and genius soared only briefly from its track. But there it was. Man had flown.

That he flew again and again, higher and farther and faster, has become the substance of an incredible half century of aviation.

Burlington has played a day-in, day-out role in that monumental saga, through triumph, failure, tragedy, and free-swinging disagreement that generally has come out in favor of progress.

The focal point of Burlington's aviation story has been its airport, which began as a cornfield and now rates with the best terminals anywhere...and which still apparently hasn't approached the end of its potential size, service, and other complexities that spell a first-class, up-to-the-minute airfield. *

In 1910, the first airplane reported in Vermont was in St. Johnsbury. Another flown by George Schmidt of Rutland crashed after dropping postcards over Rutland Fair. Those cards are now collectors' items.

---

*"Men of Vision Developed Burlington Airport -- From Cornfield to Busy Terminal", Fred Ashcraft, Burlington Free Press, December 17, 1953.

Ira Spaulding of Brattleboro made two flights in his hometown in 1913, and the late Gov. James Hartness was issued an Aero Club pilot certificate in 1916.

In World War I, the airplane became the newest of military weapons and, for the first time, they were manufactured by production methods.

The advent of the airplane generated excitement and enthusiasm, spawning the arrival of daredevils, wing-walkers, and airmail pilots.

On July 4, 1919, Mason Beebe, one of the real pioneers of aviation development, arranged for a Curtiss flying boat to land in Burlington Harbor. It stayed about a week, carrying passengers and spreading aviation enthusiasm.

The day before, John Burns and Beebe had looked over the cornfield on Williston Road in South Burlington. In a few months, the street department's horse-drawn road grader had leveled the land and a steamroller had smoothed a strip and hacked out the underbrush, and the city of Burlington rented it for $100 a year.

On August 14, 1920, a British Avro, flown by a Royal Flying Corps pilot, was the first to land at the Burlington Airport. Others had landed at Forth Ethan Allen, Colchester.

On September 21, 1921, about 3,000 people came for the formal dedication of the field. There were four planes. Gov. Hartness and Mayor J. Holmes Jackson attended.

Airmail service arrived in Boston in July 1926, and Burlington, being on the direct line from New York to Montreal was a logical place for an airmail stop. Senator Warren R. Austin began advocating to make Burlington a Port of Entry.

A Civil Aeronautics Authority in Burlington was selected by Mayor C. H. Beecher, Beebe, A.D. Butterfield, Burns, L.F. Dow, David W. Howe, H.A. Mayforth, and Hobart Spear.

In 1927, trees were cut and stumps removed from the center of the field. A landing circle was made and a wind cone erected on a pole near the entrance. Army planes visited Fort Ethan Allen during the 1927 flood. At that time, there was only an east-west runway. Norm Grady, then a high school student, remembers landing on that runway in 1928 with Freddy Borke from Middlebury. He said the field was extremely bumpy and brush had grown up.

Burlington Airport airfield 1928

The north-south runway was developed in 1928. It was 1,700 feet long and 150 feet wide. The next year, 400 feet were added on the south end. The east-west was enlarged to become 2,100 feet long and 150 feet wide. Trees were ripped down. Tractors leveled the field; runway surfaces were plowed, harrowed, seeded, and packed down. Then, fences were built.

In 1929, a Board of Airport Commissioners—Burns, Butterfield and Albert Heininger—launched work on a 80' x 60' metal hangar that had been purchased, deconstructed, and shipped in by rail.

When the pilots and mechanics complained of lack of facilities at the airport, the Street Department moved a building near the new hangar, wired it, and remodeled it for an office, pilots' room, and other purposes, including a workshop. It burned in 1931.

Butterfield suggested the city own the land rather than rent. A full-time caretaker was hired at the airport because it was getting to be a strain for Butterfield and Burns and Heininger to go to the airfield after dark to open the hangar for planes that arrived late.

By mid-1930, Curtiss-Wright airplanes were in service carrying papers and passengers between Albany and Burlington. They advertised planes would wait until 4 pm for passengers going to Albany.

Soon afterward, Nick Harris began flying passengers, parcels, and express across the lake to Plattsburgh.

Until 1933, there were gradual improvements. More land was bought and leased, markers were placed on the ends of the runways. BURLINGTON was painted on the hangar roof, more trees were cut, and a deep well and pump were installed.

The Civil Works Administration started on November 23, 1933 and a hard-surfaced runway would soon be a reality. Grading was done according to surveyors' specifications with very little help from street department equipment. Hundreds of men with shovels carried the blowing sand from the high spots to fill in the low. The north-south runway and the apron were hard surfaced in 1934, and the east-west runway in 1935. Drainage systems were also installed. A blind corner was cleared on the northeast side of the intersection and grading took place in all areas.

June 1937 to June 1938 was a big year. From City Engineer George Stanley's report we find the wooden hangar had been completed. The beacon was in operation as well as boundary and approach lights and obstruction lights installed.

The Bureau of Air Commerce spent over $100,000 on an extensive program of aids to air traffic centered around Burlington at no cost to the city. Eight airway beacons twenty miles apart were erected between White River Jct. and Burlington. Those nearest

Burlington were on Bolton, Stimson, and Robbins Mountains. The tower parts for these lights were carried by trucks, horses, and on men's backs. Beacons lighted airways from Boston to Burlington to Montreal. There would also be a new lighted route from Burlington to Albany connecting points west and New York City.

The CAA—Doc Bayne, Chet Talix, and Ben Roland—erected a radio range system with 134-foot steel towers on Winooski Heights, five miles from the airport, north of Winooski.

The receiving and transmitting instruments were installed in the airport administration building. Radio range began operating September 1, 1938, on a twenty-four-hour basis.

A 26.5' x 30' wing was built on the north end of the administration building to accommodate the radio range central station, teletype communications, customs and immigration, future airline office, and to enlarge the waiting room.

Hourly and half-hourly weather reports were broadcast.

Weather reports by teletype began March 1, 1938.

BURLINGTON AIRPORT

The Old Brick Schoolhouse, or Eldridge School, was probably built about 1865. It rested across from the southwest corner of the Eldridge cemetery. Early town meetings were held in the schoolhouse until the Town Hall was built in 1871 at the corner of Williston Road and Dorset Street in South Burlington.

Among the people to attend the old brick school were Hiram Tilley and his brothers, also Tom and Frank Mills and Nellie Mills Krupp. The teacher at the time, 1880s-1890s, was Miss Bid Redmond. Rollin Tilley went there three or four years and has a perfect attendance certificate signed by Miss Plant, his teacher.

About 1926, a new two-room schoolhouse was built a little farther west on Airport Road. The Old Brick Schoolhouse stood idle until 1932 when the South Burlington School Directors donated it to the airport as the airport had recently lost their office and shop in a fire. It was used as an office, and Harold Pugh conducted Ground School there. The fellows used it as a place to live, cook, repair engines, and wait for the weather and for business. It also served as a passenger waiting area for air taxi service from Plattsburgh or a town taxi. Civil Aeronautics Inspector, Bob Hoyt, gave examinations for private and limited commercial pilot's licenses and wrote out his ship inspection approvals from here also.

In the spring of 1933, the schoolhouse building was repaired and painted outside. Three rooms were partitioned off; linoleum was laid on some of the floors and a workshop and bench were installed. Harold Pugh did most of the work. This building was also the official headquarters of the Vermont Airplane Pilots Association. Visiting pilots declared that Burlington had the best administration building in the state.

Pilots and mechanics continued to live there. It was their base of operation as they took turns at night stoking the salamanders (oil drums on their sides), which kept the new concrete floor in the steel hangar from freezing. This was the winter of 1933-1934, and temperatures were twenty to thirty-below. Water froze in a pail only a few feet from the schoolhouse stove.

When the Boston-Maine Central Vermont Airlines started in February of 1934, the new administration building was under construction. A waiting area for passengers was sectioned off in the old schoolhouse. Schedules were scribbled on the blackboard. It was heated by the usual box stove and a stove pipe went the full length of the room. A phone was installed with an extension at the hangar. The bell out there was huge so it could be heard above airplane motors, etc. It seemed the phone always rang when the dispatcher was halfway between the two, and he couldn't decide which way to run.

In 1935 or 1936, the schoolhouse was occupied by the pie-man. He made all kinds of pies to sell, including oyster, spinach, cabbage, and onion, plus the regular varieties.

In 1940, the partitions were removed and a long table was installed for packing parachutes. Every sixty days, parachutes had to be hung up and aired to get the dampness and wrinkles out before they could be repacked.

I taught Ground School there in 1942 with a hand-operated Jam Handy slide projector, clamped to the back of a chair. The blackboard served as a frame for the makeshift screen.

We held meetings there in 1943 with our instructors and linemen as it was the only room on the airport big enough for all of them. Later it became a storage area for spare wings, struts, spars, tubing, landing gears, wheels and tires. The Civil Air Patrol held meetings there for a short time, and it was the first home of the Vermont Air National Guard when they organized in August 1946.

THE NEW ADMINISTRATION BUILDING

Building cement blocks for the new administration building

The new administration building was patterned after the model shown at the Century of Progress International Exposition in Chicago. Cement blocks for it were made at the field by Harold Pugh, George Hall, Jim Beckwith, Hoyt Gilmore, Clarence LaCasse, and others in their spare time. The building was 28' by 36'. A ticket counter was on the south-end of the waiting room. The area behind this was a shared office space for the airline and airport managers. There was a telephone, a regular Atwater-Kent radio, electric clock, and a pressure type wind velocity gauge. An electric switch control panel was located near the counter. On the waiting room wall was an airways map of the whole United States. The waiting room furniture was modernistic tubular chrome with black cushions. In the back was a kitchen with a dinette and a small bedroom and shower for the airport manager and his wife. It cost $10,787, and the work was supervised by Hobart Spear who was Airport Commissioner at that time. When Mayor Burke came on a tour of inspection, his only comment was, "Humph! A palace in a cow pasture. Good enough for a bank!" It really was nice, and we were very proud of it. We bought our own bedroom furniture, but there was no room for a chair. At City Light, we got discounts on a stove and refrigerator for the kitchen. We were expected to serve lunches to passengers, pilots, visitors, and whoever. Since I was teaching, we had to hire someone to cook and serve and keep things tidy. I helped on weekends and did the major cleaning—like washing and waxing floors and polishing windows. This was our only home until the fall of 1941 when we moved to the Lane house near the south side of the wooden hangar.

Administration lounge in the newly completed building in 1935

An iron stairway on the south side outdoors led to a flat roof with an iron railing across the front. It was a great place to watch everything including air meets, air tours, and airline activity. A chain across the bottom kept out most of the public. This same platform was used at an emergency control tower in 1942. The controllers, Clifford Ward, John Gabso, Randy Wendell, and Bill Darby, would stand on the roof with a microphone hanging out the window with a light gun and binoculars in hand directing air traffic. Inside, along the wall between the windows and tower-bracing, there was room for two cots. It was used as a bedroom by pilots, mechanics, and whoever. The center of the floor was open to the waiting room below. In cold weather the windows steamed and frosted. It melted on warmer days and ran down into the cots and froze again at night.

Air traffic controllers: on left operating a microphone and on right signaling with a light gun

The kitchen and bedroom of the administration building had been located over the cement basement of the old office building and shop that burned in 1931. This is where the hot water heating plant and water pressure system were located. There was also a tiny pot-bellied stove to heat hot water. What a monster!

A lot of trees had been cleared, and most of the wood had been piled along the south side of the cemetery fence for use of the welfare department. The wood from the trees cut south of the entrance road had been stored in the basement—crickets and all. The wood was there for several years and so were the crickets. When the basement light was turned on, the crickets would scamper away, but strange thing—they were white now, not black!

There was only a crawl space under the waiting room area. I turned the cellar light off one time then realizing that perhaps someone was down there, I turned it back on just in time to see an electrician poke his head out of the crawl space and blink, much like a woodchuck coming out of his burrow.

Administration Building and Hangars

THE FIRST BEACON

On October 26, 1935, a homemade beacon was installed at the top of the steel tower on the administration building. It could be seen from as far away as Milton, Richmond, and Mt. Philo.

Conforming to specifications obtained from the U.S. Bureau of Air Commerce, it was constructed by Kenneth Partlow, Acting Dispatcher, Harold Pugh, Airport Manager, and George Hall, Aircraft and Engine Mechanic.

They used an electric turntable from the window display of a Church Street clothing store, a large reflector lined with small mirrors, and a 500 candle-power bulb. It had a green light on the back and a red obstruction light on top. It rotated at the standard rate of five turns per minute and was tilted upward so the beam would clear the tree tops, etc. It cost about half-a-cent per hour and was operated from dark until 10 pm or as needed. If a plane came in after that, it would buzz around the airport until the manager woke up to light its way in. The switch was on the main control panel near the office. The money for a regular beacon and other lighting was appropriated in 1937.

THE WOODEN HANGAR

New wooden hangar

Construction of the wooden hangar was started in the winter of 1936-1937 and was finished late summer 1937. This Works Progress Administration (WPA) project was under the direction of George Stanley, City Engineer. It was 100' long and 80' deep and 22' high with a lean-to 80' long and 20' wide on the south end, divided into a mechanics workshop, paint shop, and office. Built of wood with no center posts, the 100' long timbers had been shipped from the west coast. There was a concrete floor and ramp.

In March, the taxpayers had authorized a $12,000 bond issue which covered the city's share, plus boundary and obstruction lights, rotating beacon, ceiling projector, and illuminated wind cone.

After the Pearl Harbor attack on December 7, 1941, severe flying restrictions were imposed on the 150-mile defense zone that stretched inland from the coast. The Northeast Airlines Pilot Training Division moved to Burlington. In 1942, they built a wing on the north side of the wooden hangar for flight operations. The Waterman Building at UVM housed the Link Trainers and classroom space. Students lived at Converse Hall. A northeast hangar was built near the place where Air North is now located (1977). It was 180' x 205' with 25-foot doors, and a 30' x 205' lean-to was added on the south side for storage, office, shop, dope room (for applying glue), stock room, and heating facilities. It was dedicated on June 15, 1943, and the twelve Stinsons equipped for blind flying, four Lockheeds, and six Link Trainers were moved in. All these were used in the instrument flying instruction for their 120 students. The school was discontinued in the spring of 1943. It was sold to Matco (Mr. Lozon) and became a factory for manufacturing aerosol insecticides. It burned on December 2, 1947.

For several years, our office was the shared space behind the airline ticket counter. Later, Harold had a desk in the corner of the waiting room. For about a year, we used the airline corner of the new addition for office space but gave it up when Canadian Colonial Airways service began. For the following year-and-a-half, we used the front room in the lean-to on the south end of the wooden hangar for our office.

In the fall of 1941, three houses on White Street were considered hazardous to the approach on the north runway and had to be moved. We acquired the Lane house with a garage, and during the night it was moved down the runway to a place south of the wooden hangar. We did a lot of work on it and used it for our home and office. We soon outgrew it and gave up our living quarters and went to live in our home at 1537 Williston Road, South Burlington.

We had purchased our home several months earlier to house some cross-country students. During the Air Force Indoctrination Program, five secretaries were needed and even our former kitchen became the office for a government inspector. Our previous office space in the hangar was used for a ready room with 200 log books and fifty-five parachutes in racks along the wall.

When the Northeast Training school moved to their own new hangar, we acquired the north lean-to on the wooden hangar where we had an office and waiting room, stock room, parachute packing room, and other.

In the spring of 1944, the Lane house was moved onto the lot right behind our house at 1537 Williston Road, South Burlington facing Woodbine Street.

FLI-RITE SCHOOL OF AVIATION AND CPT

In August 1939, Harold Pugh started the Fli-Rite School of Aviation after Congress appropriated $4,000,000 for the training of approximately 11,000 students in colleges and universities during the next year. The local training at UVM, under the direction of the Fli-Rite School of Aviation, consisted of seventy-two hours of ground school instruction and thirty-five to fifty hours of flight training at the airport. At the completion of this primary training, a trainee received a civil pilots private license which made him eligible for advanced training in the secondary course. It was an opportunity to learn to fly at an age when students learned most easily.

Grace and Harold, 1939

The school had forty-two students within the first three months of operation. Schools were also conducted in Swanton and

Plattsburgh. We had about eight groups of twenty or thirty primary students during the next three years. One group was trained to be glider or service pilots and the last group received commando training for service pilots.

Other courses were added so that we also had secondary instructors, two groups of secondary instructors (for instructing acrobatics), and four groups of cross-country-instructors. Several from Canada were also seeking instrument and other pilot ratings to make them eligible for the Royal Canadian Air Force (RCAF). Fli-Rite logged 700 hours in July of 1942.

Fli-Rite School of Aviation instructors and students

The Cross-Country Instructor course prepared trainees for a commercial license with instructors rating. They also learned instrument flying procedures but always had visual contact with the ground, preparing them for plane ferrying in government service. One student "Mac" Perkins was the first to fly the Hump (Himalayas during World War II).

There were many hours of ground school for each course. Nationally, 37,000 young men and women entered the primary program and 4,000 the secondary course. More than 4,800 of those trainees had already entered the armed services. The government regarded this program as a most vital need in our defense system.

On December 7, 1941, the Japanese attacked Pearl Harbor and immediately, by telegram, we were notified that all pilot certificates were suspended and could be reinstated only after satisfactory evidence of citizenship and loyalty, accompanied by positive identification, had been presented to the CAA Inspector of CPT supervisor. This was followed a few days later by request of airman's identification cards bearing picture, signature, and fingerprints. Because we had to meet these requirements, we didn't fly for quite a few days.

By January 1943, college enrollment was very small. CPT was no longer taking women, and most of the eligible young men were already in service.

The following year, the Air Force Indoctrination Program started. During the coming year, 8,000 of the nationwide total of 30,000 Air Force cadet trainees would be assigned in New England. They had to have had two years of college or be high school graduates with the prescribed refresher course, have passed the required physical, be a citizen of the United States, unmarried, and between the ages of twenty and twenty-four. This was largely negotiated by F.W. Shepardson, on behalf of the university and the airport. Burlington was assigned the 61st College Training Detachment.

The first group of 600 aircrewmen arrived on March 1, 1943. Under the direction of Captain Allen F. Ernst, each student was to receive thirty-seven hours of campus instruction per week plus the hours of flying. There were as many as thirteen sections of a class and the university went on a six-day schedule from 8am to 6pm. After their ten hours of dual flying instruction, the trainees would be sent to a classification center and as cadets, be assigned to training as pilots, bombardiers, and navigators.

We sold our twenty-two airplanes and were assigned thirty-two planes requisitioned by the government. Our pilots traveled 26,000 miles in six weeks, collecting them from points in New England and New York. Some had been dismantled and stored in barns. Others were tied down in pastures, but they were flown to Burlington in February and March, repaired, conditioned, and put to work.

For the previous CPT program, we had bought about a dozen nice flat Pioneer back pack parachutes. But now we were dealing with the military and they insisted on seat-packs (to bump on the back of the knees). We purchased forty-five new Switlick seat-packs costing $9,000. At last, everything was ready, and nobody had ever seen so much flying—200 students every month, ten hours each. Doc Sanders was manager of operations and chief instructor. He was the one who had to crack the whip, be negotiator, and general good-will ambassador. He was in charge of the thirty instructors. Most of them stayed the whole year, but others were subject to draft, active duty, and the lure of higher wages. There were ten or more linemen to service and handle the planes in the parking area. Hall and Avery had at least ten men in the shop, and you wouldn't believe the paperwork—200 of everything and seven or more copies of each. Reports and telegrams kept the lines busy. In the office, we had Emogene LaMire (Squeege), my good right hand for several years; Hazel Hall, super helper and sister-in-law; Fay Bell, Charlotte Marsh, Virginia Foote, and Bob Morgan—stock clerk and payroll clerk.

During the year April 1943 to April 1944, we had trained 2,600 students, 200 at a time with 25,000 flying hours.

Students

One night in mid-April 1944 with the strictest military secrecy, the students were suddenly shipped away under the cover of darkness. Leaving us with fifty-seven employees, no work, and no money for salaries or expenses.

Harold was extremely proud of the fact that Fli-Rite students had flown about 30,000 hours in the twelve years he had been at the Burlington Airport and so far, not one student had even a scratch. This was really a tremendous record, and one for which Harold and many other people should be given credit.

On December 28, 1943, when we were training the aircrew students, we were very sad when two planes collided in the Dorset Street practice area. Our instructor, Rene DeLaricheliere, and a student, Harry Hollander of New York, were killed instantly. The other plane had minor damage and no personal injuries. Rene had been flying with us for about ten years. He was a chief instructor with no regular students, but he spent several hours each day checking the students of other instructors.

Fli-Rite poster

AIRPLANES

"Among other uncertainties in those days was the probability that the engine would not start, but by certain incantations namely "contact" thundered by the pilot and echoed by the mechanics, and "spin", were sufficient to stimulate the engine into action and after a few spattering coughs, backfires, belching and blue smoke and flame from exhaust manifolds, the faithful engine settled down to its task, while the pilot, unhampered by aviation instrument panels, controls, etc. warmed the engine 'til his practiced ear said it was ready to go then pushed the throttle forward until the beat of the cylinders just matched the throbbing of his excited heart. Meanwhile, the fragile straining plane was held back by ground crew until at the signal of a raised hand, the plane lumbered and raced down the grassy field and at length became airborne."

(November 24, 1976, Springfield, *Vermont Reporter*, Adoree Allen).

AIRPLANES AT BURLINGTON AIRPORT TO 1942

Hisso WACO

Stinson, owned by W. Waite

Stinson, owned by G. Kellogg

Stinson SM8A, 420-Y, 423-M, owned by H. Pugh and 418-M (Swanton)

Stinson R

Stinson Gullwing SR9B

OX WACO

Standard

Travel Air 0XX-6

Parks KR-31

Aeromarine – Klemm

Warner WACO F

Arrow Sport Sparrow
Bird
Robin
Buhl Pup
Curtiss Pusher with Szekely engine
Challenger
Great Lakes
WACO F, 2 of them
WACO F-2
Eaglerock
WACO Cabin
Fairchild 22, 3 of them
Monocoupe
Star Cavalier with Velie engine
Pioneer Amphibian
Aeronca C3
Aeronca K
Aeronca 50, 2 or more
Aeronca Chief , 10 to 12 of them
WACO UPF-7
Travel Air 6000
Cub – several of them
Taylorcraft
Dart
Skyfarer
12 Stinsons, belonging to Northeast Pilot Training Division
4 Lockheeds, belonging to Northeast Pilot Training Division
On April 4, 1937, Burlington Airways, Inc. purchased a five-passenger Stinson Reliant. It had been on exhibition at the aircraft show in New York in February. It had a cruising speed of 150 mph, blind flying equipment, 245 horsepower Lycoming engine, controllable pitch propeller, flaps, and gull wings.

EARLY AIRPLANE OWNERS

Vermont Airways, Inc.
Queen City Flying Club, Inc.
Richard Colburn
Morgan Gillette
Nick Harris
"Spud" Murphy
Stuart Astles
Harold Taylor
Arthur Ashley
Malcolm Jones
James Beckwith
Justin Neff
George Kellogg
John Webb (J. Shaw)
Dr. A. J. Crandall
F.W. Shepardson
Willett Foster
Don and Ken Allen
Waterbury Flying Club

Fli-Rite School of Aviation airplanes at Burlington Airport

OPERATIONS

The Boston-Maine Central Vermont Airways began operating out of Burlington on February 3, 1934. It was the airways subsidiary of three New England Railroads: the Boston and Maine, the Central Vermont, and the Maine Central. The airline had previously terminated at the Barre-Montpelier Airport.

At that time, there was no radio or teletype communication. Airline arrivals and departures were sent over the regular railroad telegraph system at Essex Junction. Often the planes arrived before the message. Weather reports came by Western Union and were phoned to us. Sometimes the weather had come and gone before we got the warning.

Each morning, after we picked up our mail at the Burlington Post Office, we stopped at the weather station, which at that time was the square brick building near the present UVM tunnel. There we waited and watched the weatherman sketch for us the isobars, isotherms, fronts, lows, highs, and the usual things found on a weather map.

Mr. Milton Dow, in charge of the station, was supposed to give the forecasts for our area exactly as he received them from Albany. However, in strictest confidence and of dire necessity, he might be induced to give his own expert opinion. He boasted an extremely high percentage of accuracy. The Champlain Valley has its own weather system, and he had studied it for many years. The Weather Bureau was moved to the airport in 1943. It occupied the area previously used as kitchen and bedroom.

In the summer of 1934, the Stinson Trimotors flew in from Boston having stopped at Concord, Manchester, White River, and sometimes Montpelier. They arrived in mid-morning, and the pilots waited around the airport until afternoon, when they made the return trip. The pilots spent their time reading, swapping stories, doing their problems for the Air Force Reserves, playing golf, teasing the dog, and pitching pennies.

Customs and Immigration Offices were in the post office building on Main Street in Burlington. When the airline started making trips to Montreal in 1934, the customs and immigration officer had to come to the airport for each stop and as needed for private planes to or from Canada. The Bus Company (BRT) was tied in very closely with the airline. They made reservations, sold tickets, and provided Lincoln limousine service.

The immigration and customs officers, Luther Kennell and Frank Johnson, often rode with limousine driver Ray Bora. The airport bus service started in 1936. In the spring, Airport Road was so muddy, the driver turned the bus around at the corner of Williston Road, leaving the passengers fuming at the airport.

On February 19, 1934, President Roosevelt cancelled all airmail contracts. The army pilots, who were the first to fly the mail, were to become mailmen again, and they flew the mail from February 19 to May 10, 1934 under the direction of General Douglas

MacArthur. The army pilots did not have the supportive systems that the airlines had, and many of them were killed—in fact it was called a slaughter, and new airmail contracts were awarded.

Airlines were supposed to be in operation for ninety days before applying, but this rule could be waived due to the emergency.

Harold with the first bags of new air mail service, 1934

The Boston-Maine Central Vermont Airways (BMCV) was a branch of National Airways, Inc., and they had entered a bid of 29.5 cents per mile to carry airmail from Boston to Burlington. As this airline had been in service since early February, no waiver was needed. The first airmail flight from Boston to Burlington was on June 25, 1934.

Everyone is amused at the pictures of the BMCV Airways Stinson Trimotors with tire chains. These were the brain-child of Paul Collins, World War I pilot and president of the airline. They were specially designed so that they would not slide on icy runways.

Snowplows did their best but under certain conditions, snow packed and turned to ice—especially on the grassy field. Because they used these chains, BMCV was the only airline operating in and out of Boston for several days in February 1934.

Stinson Trimotor plane with ice chains

Air Express Service began December 3, 1934, linking up Burlington to Boston and the rest of the nation. Express could be sent to the West Indies in two days and as far as Buenos Aires, Argentina, in seven days.

Mayor James Burke, age 86 and the oldest mayor in the country, sent maple syrup to people via air express, as did several other officials.

Bill Tanner, pilot for BMCV Airways, was killed on April 8, 1936 when he took off for a test flight with only two of the three motors operating. He crashed in the trees about half a mile north of the airport. Hoyt Gilmore, A & E mechanic at Burlington, was with him. He was injured and dazed, but recovered.

In January of 1936 Milton Anderson, the chief pilot of Boston and Maine Central Vermont Airways, spoke to the Rotary Club in Montpelier. The airline had been in operation for two years, and he declared that all planes on his line would soon be equipped with two-way radios and a receiving station would be installed in Montpelier as well as transmitting stations at Boston and Burlington. He said the radio beam from Boston extended only as far north as Montpelier. These aids to navigation became a reality in the fall of 1938, when the airline had been in operation about four and a half years.

Harold Pugh left, Mayor LaGuardia center, and Willett Forster right Celebrating inaugural Canadian Colonial Airways landing at the Burlington Airport

Canadian Colonial Airways made its maiden flight into Burlington on April 1, 1939. Governor Aiken and Willett Foster, President of the Chamber of Commerce and Chairman of the Airport Commission, flew to New York to accompany Mayor LaGardia here—also included in the welcoming ceremony were Mr. Duncan and Mr. Hamburger, airline officials; Sen. Warren R. Austin; F.W. Shepardson; Mayor Dow; Congressman Chas. Plumley; Paul Collins, President of Northeast Airline and representative of the mayor of Montreal.

For at least three weeks, Canadian Colonial Airways made no other Burlington stops. They complained the runways were wet, the trees on the north end were too high, the ground at the end of the runways was rough, and the runways were too short.

The trees were cut, the ground smoothed, the runways lengthened. On September 15, 1941, they cancelled night landings because the lights were being changed. In October, the 4,300 foot diagonal had been completed and there were contact lights on all three runways. In March of 1942, they could not stop due to failure of the field lighting system. Ten times in the last four months, they said, smudge pots had hurriedly been set out. This caused passengers on Canadian Colonial to be unwillingly taken to Montreal or Albany or New York.

Trucks working on the runways had damaged the wires and breaks were common. A whole new system of runway lights was installed in 1946.

Yankee Skylines was incorporated in the State of Delaware on June 29, 1944. It was the brain-child of Harold Pugh, John Adams Browne, and several others. The plan was for an east-west scheduled feeder system, including Ogdensburg, NY, Malone, NY, Plattsburgh, NY, Portland, ME, Bangor, ME, and Concord, NH. It would connect these places with the regular Northeast and Colonial Airline stops. A lot of work was done, and there was a filing cabinet full of papers and letters from important and interested people. The only thing lacking were planes and operating money. Nobody seemed to have any. Corporation stock, etc. was much too complicated for me to keep track of, and it looked like an old man's paradise with goodness-knows-who paying the bills.

In 1948, Bob Noorduyn, an early aviation pioneer, was designer and builder of the Noorduyn Norseman airplane. He became president of Yankee Skylines, Inc. He went before the Civil Aeronautics Board (CAB) requesting three routes converging at Rutland. These routes would have stops covering upper New York, New Hampshire, Maine, and Vermont.

Most of those stops are now (in 1977) covered by Air North, Allegheney Commuters, Air New England, and Delta.

AIRLINE DISPATCHERS

Chief Airline Dispatcher for Boston-Maine Central Vermont Airways was Point Littlefield.

LOCAL DISPATCHERS

Harold Pugh
Ken Partlow
Harvey stone
Bob Ward
Art White—Colonial
Mike Shaughnessey
James McGuire—Colonial
Steve Waterman
Red Dow
Randy Pearsall

RELIEF DISPATCHERS

Bill Eaton
Warren Smith
Tiny Rollins
Douglas Philbrook

Harvey Stone

EARLY B AND M PILOTS

Chief Milton Anderson—Division Director 1942
Hazen Bean
Bill Tanner
Stafford Short
Don Stuart
Charles Emmerson
Kendall
John Patterson
Freddy Lord
Sam Chandler

FLYING CLUBS

Vermont Airways was an early aviation company based at Newport, with members Walter Cleveland, Russ Merriman, and Lt. Lyle Churchill. They were promoting aviation in general and selling steel hangars. They sold three hangars, one each to Burlington, Montpelier, and Springfield. Lt. Lyle Churchill represented them as the first fixed based operator at Burlington.

The Queen City Flying Club was incorporated on December 23, 1930, with Francis Clark, Raeburn McMahon, and Henry Isham, all of Morrisville. Their first plane, a 100 horsepower Kinner Bird on skis, cracked-up on Lake Elmore. They reorganized with Chas Copp, President; G.R. Collins, Secretary; Elliott Cameron, Treasurer; and Stuart Astles, Vice-President. They bought a new Parks P2 with an Axelson motor, and Bob St. Jock flew it back in April of 1931. That same summer, Harold Brannick and Steward Marcott took off in a Laird plane, which had recently been repaired but not inspected. At about 100 feet, the wing buckled, and they crashed at the south end of the field near Williston Road. Both were killed. Stuart Astles, before he had time to lose his nerve, jumped into the Parks P2 and took off, forgetting to turn on the gas. When the engine died, he turned back toward the field and crashed. That ended the Queen City Flying Club.

The Waterbury Flying Club owned a Wright J6 Waco. Walton Waite was pilot and instructor. The Eagle Rock was owned by Don and Ken Allen. Don soloed on August 9, 1933. A Star Cavalier was owned by Don Allen and Ken Tomlinson.

Ken Austin was chief participant and promoter of the Champlain Flying Club. Members at different times included Art Hinckley, Jim Clarey, Giles Tinker, Bernard Zais, Paul Chausse, Jimmy Quinn, Camille Beytebiere, Bill Mason, Lt Donald Armstrong, O. Rodney Levin, Thomas Dietrick, Allen Gray and others.

In February of 1942, Civil Air Patrol for Vermont was being organized under F.W. Shepardson, Wing Commander. That same summer meetings were held in the ready room of the wooden hangar. Bill Mason was First Aid Instructor; Art Hinckley, Navigation; and Sgt Landa Military Courtesy and Discipline. Maj William Mason became Wing Commander later.

CIVIL AVIATION AUTHORITY INSPECTORS

Bob Hoyt
Dome Harwood
Jack Summer
Royce Kunze
Gilbert Smith
Carl Rothenberger, CPT.
Tom Davis
Edgar Huestis
Glenn Jone
Tom Gates

Ken Murray
Frank Boester
Paul Miller
Everett Chase
Jack Rohrer
Robert E. Thompson
Stephen Bistran
Frank Borys

INSTRUCTORS – CIVIL PILOT TRAINING AND OTHERS

Harold Pugh
Walton Waite
Harold Taylor
Giles Tinker
 John Webb (J. Shaw)
Art Hinckley
Anne Cook
Romey Yates
Ken Tomlinson
Robert St. Jock
James Beckwith
Curley Mullaney
Sepp Ruschp
Rene DeLaricheliere
Harry Webb

ADVANCE FLYING INSTRUCTORS INCLUDING CROSS COUNTRY

Giles Tinker
Harold Taylor
Dolph Brelia
Romey Yates

Ken Tomlinson

DESIGNATED FLYING INSTRUCTORS' EXAMINERS

Harold Pugh and Harold Taylor

FLI-RITE SCHOOL INSTRUCTORS

John Bacon
Walter Bell
Marshall Bentley
Robert Birnbaum
John Byce
Marcel Chevalier
Emile Couture
Philip Datnoff
John Dockendorf
Paul Davis
Charles Hawkins
Gervase Healey
Robert Hill
Donald LaCouture
Floyd Letourneau
Melvin Luck
Robert Martin
George McDonald
Erwin Mellor
Ray Mosteller
Peter Pespini
Walter Parker
Francis Robinson
Sepp Ruschp
Wallace Sanders

Edward Steen
John Sullivan
James Tomasi
Kenneth Tomlinson
Ralph Towle
Stephen Waterman
Maurice Williams
Henry Winslow

BARNSTORMING & AIR MEETS

The pilots, or barnstormers, would fly from town to town, landing in farmers' fields or wherever they could, and then the pilot would take passengers for a ride in his plane. That was the way most people got their first airplane ride. People paid 25 or 50 cents just to see a plane get off the ground and circle a few times. They paid another dollar or two for a ride of ten minutes.

In 1933, a cooperative venture among the pilots, the air meet, was launched. Each weekend, weather permitting, a meet would be held at some airport, the location being rotated. The meets were intended to promote interest in flying, match pilot skills in competition, and make some "gas money" by flying passengers. The entire group would share the expense of advertising and staging programs. Meets were held at Burlington, Morrisville, Montpelier-Barre, Swanton, Newport, Malone, Lake Placid, St. Johnsbury, and White River.

These events at other locations worked very well, especially in the summer of 1934 when our field was so torn up that an air meet was impossible.

It would have been hazardous if stunts and races were conducted when passengers were being taken up or other pilots or airlines needed to land. Each field had white cloth fastened to a set of frames so that near the center of the field the white-cloth-on-frames could be arranged in a very large C or O meaning "closed" or "open" to other than the performers. The field was never closed for very long at a time.

The last few meets were well attended, but the riders were few. Probably all those who were interested had already had a ride. Also, the regular traffic was annoyed and annoying, so we didn't have any more at Burlington. That phase of flying had passed.

Prizes in the various kinds of competition included a carton of cigarettes, a case of oil, a sport shirt, bathroom scales, a clock, toilet kits, and scarves.

A ticket seller or two always went with the pilot to the air meet. When he made a sale, he tore off a large ticket for the passenger to keep as a souvenir and held the money and book of stubs until it came time to settle up. Some of our best ticket sellers were Hoyt Gilmore, Pat Morris and Rene DeLaricheliere, Eddie Shimeld, Allen Reed, Jim Allard, and Don Boardman.

Entertainment at an air meet consisted of:

Wing-overs

Slow rolls

Snap rolls

Loops

Side slips

Spins

Upside down flying

Air races

Parachute jumps

Dead stick landings

Bomb dropping

Paper cutting

Balloon bursting

And crazy flying such as when a pilot, dressed up like a woman, accidentally takes off. The plane does crazy things while the woman finds a book about how to fly but then loses it in the prop wash.

One celebrated air meet in Burlington in 1931 featured the pilots doing their usual barrel rolls, loop-the-loops, and dead stick landings, thrilling spectators. I remember that meet and was amused by the printed program for the affair. It carried advertising that was paid for by undertakers.

AIR MEETS

September 21, 1921: Formal dedication, 3,000 people attended, four planes, Wing walker—Curtis Oriole

September 3, 1928: Field dedicated, Bank from Fort Ethan Allen

Memorial Day, 1932

July 9 and 10, 1932: Betty Lund and Lem Povey

September 18, 1932: Four day meet, Ford Trimotor

July 3 and 4, 1933: Maj. Jack Sterling

June 16, 1935: Harold Johnson—jumper

November 30, 1935: Light Airplane Club from Montreal, six planes, fourteen passengers

August 30, 1936

June 24, 193: Arnold Ingalls—jumper

PLANES AND PILOTS INVOLVED IN AIRMEETS AT BURLINGTON

Lt. Lyle Churchill: 5-place Standard and Stinson Cabin (Kellogg's) Plattsburgh

F.E. Buswell: Rutland, WACO

Capt. Everett Marden: Middlebury, Stinson and 4-place Standard

Ralph Stancliffe: Swanton, Warner Commandaire

Dan Wilder: Swanton, OX WACO

Dolph Belia: Aero Marine Klemm, Keesville

Emery Denis: Montpelier, WACO
R.C. Drugg: Springfield, Fleet Trainer Challenger—Robin
P.E. Zimmerman: Westport, N.Y., Bellanca 3-place
Harold Pugh: WACO F or F2 or Stinson
C.V. Cressey: Concord, N.H., Kinner—Bird
Walt Waite: Stinson Cabin, Brattleboro
Robert St. Jock: OX WACO, WACO F2, Arrow Sport
Slim Emerson: Ford Trimotor
Miss Albany from Mohawk Airways
Phil Iselin: whatever available
Peter Murphy: 5-place Fairchild Cabin
James Beckwith: WACO 10
Robert Neff: Fort Ethan Allen, Challenger C-2

PARACHUTE JUMPERS

Bob Keating made 150 plus jumps
Harold Johnson
Arnold Ingalls
Ripcord Russell
Bob Ward (airline dispatcher)

PARACHUTE PACKERS

Dolph Brelia
Ken Tomlinson
Slim Hedman
Emogene LaMire

MILITARY

The most exciting thing to happen at the airport in 1936 was the winter maneuvers, from February 5th to February 15th. They were designed to ascertain the effectiveness of Air Corps personnel under the severest of northeastern conditions, using available standard and special winter equipment.

A group of twenty P26 A's from Barkdale Field, LA, were located at Burlington, and a squadron of Martin Bombers from Langley Field, VA, were stationed at Concord, NH. Eighteen attack planes were also involved.

The War Games resulted in newspaper headlines saying, "Army Bombers Destroy City," "Burlington a Mass of Ruins," and "Burlington Airport Completely Destroyed" for the third time.

They were also seeking solutions to other problems, such as field shelter for Air Corps units operating where no housing facilities were available in sub-zero weather. Waterproof tents, down sleeping bags, pneumatic mattresses, fleece and fur-lined clothing, engine hoods, and special engine heaters—all were used and tested. Officers were quartered at Fort Ethan Allen. Some oil reserves and equipment were stored in caves dug in huge snowbanks.

A new rotary type snow-thrower, called a Sno-go, demonstrated how runways and snowbanks could be cleared.

There were two accidents—no casualties. The first occurred when two planes were taking off at practically the same time, with high snowbanks on either side. The swirling snow from the first plane blinded the second pilot, who then hooked into the snowbank.

The second accident happened to a pilot from Louisiana, who had been on dawn patrol for two or more hours in sub-zero weather. Numb with cold, he plunged into a snowbank after landing.

Several types of military aircraft stopped during this time, with officers and other personnel and equipment; one was a large Douglas transport which was the first flagship of the air. It had been transformed into an aerial headquarters from which staff officers might direct fighting planes in combat. It had two radio transmitters, no armament, and could climb to 23,000 feet, which would be out of reach of anti-aircraft guns. Besides the radio-room, there was the staff room with space for eight people to work, a lunchroom, a lavatory, luggage space, and a private office for the commanding officer, who at that time was Major-General Frank M. Andrews, Air Corps Chief. The cost was $85,000, and two more were on order. We thought it was so-o-o wonderful.

One P26 A had wheels protruding down through its skis for either snow or runway landings.

Several observation planes and others made flights to the field. One aircraft had the communications equipment. The two main radio systems were for "ground to air" and for "point to point" such as weather, arrivals, departures, and requests for supplies and equipment. Telephone, teletype, and telegraph were the back-up systems.

The weather was most cooperative as we had the usual variety of conditions, and the temperature stayed well below zero most of the time. However, the next few winters, they maneuvered on Lake Michigan. Vermont was too cold!

CELEBRITIES

Lincoln Beachey flew a Curtiss Pusher in Vermont about 1911. I have a picture of him flying the airplane at Middlebury Fair. It was taken by my father who was photographer for the fair at that time. John Burns said Beachey landed at Fort Ethan Allen. He also went to Plattsburgh where George Kellogg rode with him. Lincoln Beachey was doing stunts from the catwalk of a dirigible in 1905. He became a dirigible pilot but in 1910 changed over to an airplane. He was instructed by Curtiss and became a stunt pilot—racing and winning over Barney Oldfield in his car. He flew over Niagara Falls. He was killed stunt-flying in California in March 1915 at the age twenty-eight.

Hugh Finnegan's scrapbook had an article saying Mrs. Francis B. Tomkins, daughter of Col. Herbert Johnson (of Camp Johnson), was the first woman passenger in the state in 1915. Pilot George Gray had flown a Wright pusher biplane from Plattsburgh to a state reservation near Fort Ethan Allen. Gov. Gates and others were too heavy. She said they got up to 500 feet and came down very smoothly. A letter to Mrs. Tomkins from the Aero Club of America states that in 1915, the US Army and Navy have together less than twenty airplanes available and only half a dozen of the US licensed aviators made flights of more than fifty miles.

The highlight for the year in 1931 was the American Legion Air meet on August 15th and 16th, featuring Wiley Post and navigator Harold Gatty, round-the-world flyers. There was record attendance and a galaxy of world-famous flyers including Captain Frank Hawkes, Briton and Bayles, Miss Maud Tait, and C.A. Nott. Hawkes was a three-time transcontinental record breaker in 1929, and Maud Tait, with her Gee Bee, was a Cleveland Air Race winner. Lowell Bayles had won the Bendix Trophy.

Maud Tait, famous woman pilot, and Frank Hawkes, WW I aviator, barnstormer, and speed record pilot, stopped in again during late summer 1932. They were having some difficulties with customs. Maud was piloting a Gee Bee speedster, and Frank was flying the Texaco Travel Air Mystery Ship in which he had made many records.

Betty Lund, who came to the airport for the July 9th and 10th 1932 air meet, was known as the world's most famous woman acrobatic stunt flyer. She was also a contender in the Cleveland speed races. She was well known for her daring outside loops and had recently been doing moving picture work. She was only twenty years old, and the wife of the late Frederick Lund, a noted stunt pilot who was killed in Lexington, KY, the previous October.

Performing for the same air meet was Upside Down Lem Povey, New England's outstanding stunt aviator. Several years later, he organized the Cuban Air Force.

Major Jack Sterling was the main attraction at the July 4, 1933 air meet. As a member of the Royal Flying Corps, he had made the most trips across the English-channel. Flying a small WACO, he performed many of the World War I "M" maneuvers, including falling-leaf, Immelman turn, and French reversement.

Wiley Post was honored at another air meet when he made his second visit to Burlington on August 16, 1933. His tour was sponsored by Socony. After speeches in the park, he rode at the head of the parade in an old-fashioned barouche drawn by two white horses. He was guest of honor at a large luncheon and toured the orphanages. He had recently (June 15—June 22) circled the globe a second time in 7 days, 18 hours, 49.5 minutes covering 15,596 miles. He had broken his own record made in 1931 by nearly a whole day. On August 17, 1935, Post and Will Rogers, the famous humorist, on a tour of Alaska, crashed in the fog fifty miles outside of Fairbanks and were killed. Harold flew our plane, draped in black, over Burlington during the hour of the funeral.

Capt. William H. Wincapaw, who had been flying since 1911, brought the La Touraine Coffee Ship into Burlington May 7, 1934. It was a large Travel Air monoplane powered by a 420-horse power wasp-engine. He gave rides to coffee dealers as an advertising campaign. The left wheel or shock absorber was damaged on take-off. Harold and H. Spear ran out on the field holding up a tire. He got the idea and landed on the Isham farm where the ground was softer. There was little damage.

Amelia Earhart, airline official and long-time friend of Airline President Paul Collins, was here in May of 1934 to help inaugurate the airline. May 21st was proclaimed "Amelia Earhart Day." It was the second anniversary of her famous solo flight across the Atlantic—Newfoundland to Ireland in 15 hr. 18 min. Eleven-year-old Jack Shearer carried the key to the city on a maroon-colored velvet pillow. It was presented to Miss Earhart by the mayor. In the evening, she was guest of honor at a dinner in the Hotel Van Ness, given jointly by the Rotary, Lions, Zonta, Klifa, and Athena Clubs. The next morning, May 22nd, she was given a reception by the schools of the city at the Memorial Auditorium. She did not fly the plane herself, but acting as a most charming hostess, she rode on each flight as a hundred women, several girl scouts, and others (seventeen flights and 158 persons in all) were taken on courtesy flights over the city of Burlington. Five of the women had won the trip as a prize for guessing the number of beans in a jar at the bus terminal. Mayor Burke and seven assistants counted the beans. Among the passengers to take a trip over the city was the eighty-six-year-old mayor. He had a long-standing resolve that he would never fly and steadfastly turned down Wiley Post and many others. However, Miss Earhart charmed him as she had charmed everybody. At the end of the flight, he addressed all the airport spectators crowded inside the roped-off area with the words, "If not, why not."

Those who heard her speak or came in personal contact with Amelia Earhart had to agree that the courage and skill as America's leading aviatrix were exceeded only by her charming personality.

She had an engagement in Pittsburgh, PA, that night. *The Burlington Free Press* reported that she, waving a hasty yet smiling farewell, clambered aboard the fast red-winged WACO biplane of Airport Manager Harold Pugh shortly before three o'clock and left Burlington where she had been a guest for two days.

Amelia Earhart with Harold

Manager Pugh telephoned from Albany one hour and twenty-five minutes later that they had arrived safely in the New York state capital after passing through two thunderstorms and that Miss Earhart had caught her train with time to spare. He returned to Burlington in the fast time of fifty-five minutes, announcing that the foremost woman flier had complimented him on his knowledge of the territory between here and Albany. Amelia Earhart sent a telegram saying, "Harold, I am sure awfully glad you were 'on the stick' this time and not me."

Harold was paid $36.70 for the flight by Boston-Maine Central Vermont Airways.

Amelia Earhart Putnam made another short stop at Burlington Airport on March 7, 1935 while on her way to speak before a record crowd at the State House in Montpelier. She asked the legislature to promote aviation by appropriating money to improve air fields and facilities. She told them that they might think an ox-cart was safer than an automobile, but she hadn't seen any oxen tied up outside the State House.

On January 12,1935, she had become even more famous by flying solo from Honolulu to California, 2,408 miles. She had flown the Atlantic and the Pacific!

She had also won recognition on August 24 and 25, 1932 when she had flown non-stop from Los Angeles to Newark in 19 hours 5 minutes, setting the women's long- distance record of 2,448 miles.

Hugh Herndon, who had flown both the Atlantic and Pacific, arrived at Burlington in a Stinson monoplane to address the Rotary Club. He was making a tour of 200 airports to demonstrate the practical business man's use of air transportation. After crossing the Atlantic, he and Clyde Pangborn flew from Japan to Seattle in forty-four hours, the longest over water flight in history (1933).

Jack Wright spoke before the Rotary Club on May 13, 1935. He had flown with Clyde Pangborn in an air derby from London, England to Melbourne, Australia. In reference to the Burlington Airport he said, "In comparison with the population and terrain, there is not a better airport in the country."

Lowell Thomas and Lowell Thomas, Jr. were frequent visitors. They came by airline to ski in Stowe and often had coffee in our dinette.

Sonja Henie, 1932 Olympic figure skating champion, stopped in with her mother, manager (brother-in-law) and others. They were clearing customs, etc., and stayed a few hours. She had also made several movies. The boys all said she had pretty knees and a beautiful mink coat.

In June 1935, an unexpected feature of the air meet here, was the arrival of three US Navy planes from Squantum, MA. They demonstrated formation flying, stayed about three hours and left.

Captain Albert W. Stevens arrived in a Faulkner C-14 transport monoplane in the spring of 1936. He and Orville Anderson had broken all previous altitude records by ascending to 72,395 feet in a stratosphere balloon on November 11, 1935 at Cedar Falls, South Dakota.

Princess Wilhelmina and Prince Bernhard of the Netherlands came here via Montreal to ski at Stowe.

On February 14, 1940, Doris Duke Cromwell, heir to tobacco millions, made a short stop. While Canadian Colonial stopped here for minor repairs, she chartered a local plane and Harold flew her to Montreal for connections with Ottawa.

Four bombers on maneuvers were stationed here in late August 1940. Four pilots and six other servicemen were involved.

THE PLATTSBURGH FIELD

The Plattsburgh field was at the north end of town between the railroad track and Route 9 where it ran very close to the lake.

When Harold was learning to fly there, it was called Mobodo Airport—leased from Wilfred Crete.

There were quite a few students and several planes on the field, including a D25 Standard Trainer, Stinson, Challenger KR34, and an Aeromarine Klemm. Nick Harris had started a scheduled airline service to Burlington. Service ceased after a crash at Burlington Airport in foggy weather. George Kellogg began the Plattsburgh Air Transportation Co. Instructors included Lt. Lyle Churchill, Frank Buswell, and Henry Duffany. Maintenance work was done by Buswell and Hoyt Gilmore. There was a metal hangar and an office building. We did charter work from Burlington to Plattsburgh in 1936. Jim Beckwith was an instructor there for a while.

We got a contract in 1937 with the *Burlington Free Press* to carry newspapers from Burlington to Plattsburgh Airport in order to meet the train to Lake Placid every morning.

Lloyd Hazen was an instructor for a year or so before joining the airline. He and his family lived in the office building. Other instructors were Walton Waite, Walter Bell, George Roberts, Floyd Stickney, Lee Bouyea, and Dolph Brelia. Maintenance work was done at the Burlington Airport.

A V5 Navy CPT program was carried on with Fli-Rite School of Aviation in conjunction with Plattsburgh State Normal School. It was a busy place.

In July 1941, there was an air meet—the first since Post and Gatty were there ten years before. Several planes participated, mostly ours, and Ripcord Russell was jumper.

A few days later, the wind ripped off the sheet-iron roof of the hangar, rolling it up in the shape of a ball and laying it on the ground not far from the hangar. Some of the metal was rolled into a many-layered cone and was still on the open roof. Two airplanes in the hangar had changed places but were only slightly damaged—one with a bent rudder and the other a broken cowling. A Navy plane, tied outside, broke its ropes and was lifted into the air twice. One of the radio crew was able to hold it on an even keel, and when the wind abated, it was only a few feet from its original position.

In August 1941, discussion centered around a proposed new airport as $386,000 was available in government funds. There was the possibility of passenger and airmail stops as well as a possible base for military planes.

Harold and Mayor Bouyea worked many, many hours on this new project making plans, filing forms, and laying out the new field. When it was done, there was a "P" in the center. Mayor Bouyea declared the "P" was more for Pugh than Plattsburgh. Each of the three runways was 5,000 feet long.

I don't know when the old field folded or any of the details concerning it. In 1977, there was a large trucking concern in the corner once occupied by the hangar and office.

The new airport was renamed the Clinton County Airport.

A year later, the Plattsburgh Bomber Base, costing $36,261,000 was approved. A new 8,000 foot north-south runway was constructed between Clinton County Airport and the old Plattsburgh barracks. The bombers were very heavy, and they practiced with heavy loads of fuel. Their approaches and takeoffs needed to be very flat and were a hazard to other planes. They had their own control tower, which had to be contacted before venturing into that area. The Bomber Base was one more reason for locating the radar station in Burlington.

Sign located at Missisquoi Airport

The Missisquoi Airport was started in 1928 by the Missisquoi Airport Corp. and was organized by local businessmen and flying enthusiasts. In 1934, it was leased to the Village of Swanton and with the aid of a Federal grant, the field was expanded in the shape of a triangle with a 2,000-foot runway. There was natural drainage and no hills in the area. There was a metal hangar for six planes, and an office building was erected. It was one of six border airports in the United States designated as a Port of Entry. Ralph

Stancliffe operated for a few years, and there were occasional air meets, such as Old Home Day on June 23, 1938. There was very little activity due to requirements of the customs and immigration offices. Men from the Highgate office came down when called.

Harold Taylor was instructor there in the middle to late 1930s. Ray Kimball was an active airport promoter.

In July of 1940, activity began to pick up. There was an air meet with Ripcord Russell, parachute jumper. At that time, the airport was chosen, along with Springfield and St. Johnsbury, for a non-college CPT program. Fifty students were to take ground school. Five students with the highest rating would be given thirty-five to fifty hours of flying time for a private license. The program was free of charge except for their physical, which cost $10.

On May 29, 1941, Fli-Rite was CAA approved and received air agency certificate #1561.

Phil Iselin was the instructor for the first course. Fred Kelley started the second course, but Dolph Brelia finished the course.

I drove to St. Albans two nights a week and taught meteorology and helped with some of the other ground school. On May 29th and 30th of 1941, there was an air meet with Bob Ward as parachute jumper. Fred Kelley crashed on the field doing a crazy-flying act. He died three weeks later.

About the same time, Walter Bell, Roger Chevalier, Lester McCarthy, Ray Wood, Ray Kimball, and Rupert Blatchley formed the Missisquoi flying club. By July 1941, Marcel "Sam" Chevalier, age sixteen, had seventeen hours solo but had to wait until age eighteen before getting his private pilot's license. He continued to build flying time and kept things going when nobody else was around. Dolph Brelia came when there was instructing to do, and Harold was there every few days.

We stopped operating there that fall as we needed the planes and instructors at Burlington. In April 1942, St. Albans and Swanton discussed owning the airport jointly.

Two years later, our indoctrination program was at an end, and again we set up a flying school under Arthur Viens. Harold Taylor also helped with the instruction, and later Earl Smith was in charge of the shop, which had been built on the back of the hangar.

There were plans for several small hangars. A ground school was in progress, and students were flying on the GI Bill. A special trailer truck would carry a plane to the repair shop from any part of New England.

There was a big air meet in the spring of 1946, as Mr. Way, owner of the airport, renamed the field "Austin Airport" after Senator Warren R. Austin, whose birthplace was in nearby Highgate.

In June of 1946, we sold out to E.W. Coe, formerly from Ogdensburg, NY.

Six planes were located at the field with accommodations for nine. There were several T-hangars. Our old pal Earl Smith was manager, instructor, ground instructor and repairman. In 1969, the state purchased the Austin Airport from Chester Way. I believe the field was relocated soon after because it was so near the school. The next year, a hard-surface landing strip 3,000 feet long and 60 feet wide was installed. To protect the new Missisquoi Valley Union High School, the direction was slightly changed from the old grass runway.

I was there in 1972, and there was nothing but a huge corn field on the site of the old airport. The new airport was on a side road and north from its original site. There was a hard surface runway parallel to the school, two hangars, and a dozen or more planes.

I REMEMBER

During the summer and fall of 1937, we had early flights each morning delivering bundles of *Burlington Free Press* newspapers to Plattsburgh Airport. The newspapers were placed on board D&H Train No. 7 at 5:22 am bound for Lake Placid, NY. In those days, this was the only way the Associated Press news would be delivered in a timely fashion to the Lake Placid area. Each morning at 2 am we would check with the Burlington Weather Bureau. If it looked like poor flying weather, we would take the newspapers by car via Champlain Bridge to Port Henry and then load them aboard the train at 3:42 am. Of course, Jo-Jo, our dog, always went too, whether we went by plane or by car.

Initially, we started with about 100 pounds of papers and we could easily use the Fairchild 22. In a short time, however, there was about 400 pounds and we had to use the Stinson. According to the contract, the pay was $5 per day, and there must always be a spare pilot and plane. Early morning fog was the greatest hazard.

Another time, Harold and George were hurrying to beat the train in Port Henry, and as they rounded a curve not far from the West Addison church, a whole flock of sheep was taking advantage of the warm asphalt by sleeping on the foggy highway. All were safe.

* * *

We always had to be alert for dangers. One time after someone had been cutting hay on the field in Plattsburgh, a hay loader was left in the middle of the runway. Harold was startled when he saw the hay loader through the fog as he was about to land.

***

That same fall in 1937, I was teaching in West Panton, a mile or two from the Champlain Bridge. One Monday morning after we had made a speedy run to Port Henry, we stopped by my boarding place and slept in the car until it was time to go in for breakfast and then to school.

***

There was a large wheat field behind my schoolhouse. One Friday afternoon after the wheat had been cut, Harold flew down in the little yellow Aeronca to get me. It was a warm day, and the windows were open and there were no screens. Harold was always early, and he got there before school was out. When the kids heard the airplane so close, they didn't bother with the door—they went right out through the windows. Jack Smith, who had worked in the Control Center at Nashua with Tom Shimeld, was one of the kids. He told the same story.

***

Harold's daughter, Barbara Pugh, spent at least three summers with us and made herself extremely useful. She was loads of fun, too. I enjoyed joking with her. One afternoon, she had finished her work in the office at the hangar and wanted to know what she could do next. I couldn't think of anything right then, but I told her to get a bucket of prop-wash. She went over to the administration building. In the basement, she found a bucket but no prop-wash. Not knowing exactly what she was looking for, she inquired of everyone she met. Amused expressions and confusing answers added to her dilemma until she met Loren Avery. He told her there was some prop-wash going down the runway behind that Aeronca that was just taking off. Then she came back to me! I was extremely embarrassed to tell her I had tricked her—it was a part of airport initiation.

The first summer Barbara was at the airport, she was about fifteen. She had learned to do the Fli-Rite paperwork and was useful in other ways. I went to New Jersey and Pennsylvania with Harold's sister for about eight or nine days. Meantime, the gal who ran the kitchen decided to go home for a few days for a birthday. This left Barbara with the kitchen and all the guys to feed, plus the cleaning and the office work. She hustled around and did a good job.

* * *

Vermont Airplane Pilots Association Meetings were held on the apron in front of the administration building. Lights from the waiting room, the overhand, and the beacon were sufficient.

* * *

I often cleaned and dusted the waiting room. As soon as I would finish, some airplane would turn around outside and blast a lot of sand and dust in through the open door, and I would have to clean again.

* * *

Rollin Tilley, a local farmer who brought the milk every morning, and would lean on the doorway and listen as the boys did their early morning hangar-flying.

* * *

Bill Ashline, engineer for WPA, would get a phone call about every five minutes.

* * *

Harold was designated CAA flight examiner on November 7, 1942. At that time, he had a total of 3,692 hours in the air, 2,600 hours as instructor. By April 1941, he had carried 12,200 passengers and had stopped counting. He had held his license for about ten years at that time. He carried the passengers one or two or three at a time, adding only about ten minutes to his log book each time. Most of the student instruction was for half-hour periods. His logbook resembled a diary with students' names, the kind of practice (as spins, eights, power turns, stalls), and charter trips. He even mentioned the destination and stops.

* * *

Harold was better known as Airport Manager, but his greatest accomplishment and contribution was his flying school. It was one of the best in the Northeast, and his graduates were a source of pride wherever they went. Many received outstanding recognition in various parts of the world.

* * *

Walking down to get groceries at John Bradley's store at the Y where White Street meets Williston Road, the track was narrow, the snowbanks were high, and the drifts were deep. There were about six houses on the north side and a barbed wire fence. It sure was cold!

* * *

Jo-Jo and Barnes' dog hated each other. When the airline was about to take off, the noise seemed to infuriate them. They would fly at each other and have a fight near the runway, sometimes delaying departure until the dogs could be returned to their respective homes.

* * *

Airline pilots and dispatchers would often be dressed properly for dinners and luncheons. Their uniform was a navy-blue suit, light-blue shirts, and caps with black visors.

* * *

When handling mail or meeting planes with airmail, it was required to wear a hip holster with a revolver. Sometimes when Harold was flying, I had to meet the plane. I wore the holster, but I would never have been able to take the revolver out—much less shoot it. I also had to wear the hat, which was much too big. If I leaned forward it fell down on my nose, so I wore it sideways!

* * *

One time, Pilot Bill Tanner had to deputize a passenger so they could hold the revolver on an unruly intoxicated person until they could land.

* * *

On October 19, 1939, a pilot on his way from Boston to Augusta, Maine, became lost in the darkness. He flew over the village of Lyndonville and, cutting back on his motor, shouted for help to direct him to an airport. Trefren and Wood, two undertakers, took their ambulance with the spotlight as a vertical beacon and, with the siren wide open, drove five miles to the St. Johnsbury Airport while the aviator followed in the sky. At the airport, other automobiles circled the field and furnished headlights for a safe landing.

* * *

Norm Grady lived next door to George Schmidt's mother in Rutland when he was a boy. He remembers seeing the wreckage of the plane (1913) in her barn, at the corner of Gibson Avenue and Granger Street.

* * *

Four fellows were living in the Old Brick Schoolhouse in the winter of 1933. They were hungry, but the roads were bad so they walked to the store about a mile away. They looked around and decided they all liked rice. "I can eat that much," said one, holding up a package. "So can I," said each of the others. They went back to cook the four packages of rice on the old box stove. They emptied all the packages in one pan and added water—enough to cover. Soon they added more water, and then they had to get a larger pan and more water and a still larger pan. Finally, they had to throw the excess out the door so they could cook enough to eat.

* * *

One night, Harold, Rollin Tilley, and I flew up in the Stinson to watch the sunset. We saw the sun go down behind the mountains several times as we climbed to about 7,000 feet. It was beautiful as red and gold streamed through the clouds on the horizon. We returned to the field immediately, but I was startled to find the airport was so dark. I didn't know where we were landing.

* * *

Harold would have to assist stray pilots who needed to land after dark. When we would hear an airplane engine overhead, he would have to get out of bed, drive to the end of the runway, and shine his headlights so the pilot could land safely.

* * *

Harold Taylor's student, Hiram Fernando Evans, was taking off toward the west. Taylor ran out toward the runway near the cemetery and tried to tell him with motions that he was taking off downwind. "What?" said the student, pulling back on the throttle and sticking his head out the window. From that day on, the student was always called "Hiram Fernando Down-Wind-Throttle-Back Evans."

* * *

Our former student, Dr. A.J. Crandall, who owned a Monocoupe, left a large surgical and medical practice in Essex Junction to enlist for overseas duty. Maj. Crandall and his famous auxiliary surgical unit landed by glider in France behind enemy lines two days before D-Day. There, they set up a hospital in a French chateau to care for the wounded. A captured German medical unit assisted them.

* * *

Rollin Tilley, who had a ride in the Curtiss Pusher, said, "It felt like I was floating around on a goose egg." He also remembers cutting the hay on the airport in 1932 or 1933 using horses.

* * *

Secondary Instructor Course students had to learn to do chandelles and lazy-eights two different ways because they never knew which inspector might test them—the one trained by the Army or the one trained by the Navy.

* * *

Mr. Webb's eighteen-year-old son, J. Watson Webb, Jr., located in Boston, had been stricken with acute appendicitis. Dr. Foster had suggested that Harold fly J. Watson Webb to Boston. A night flight in those days was hazardous. They were in the WACO F2 with open cockpits, no lights for the instrument panel, no airway beacons or markers, no radio, and plenty of mountains. The high flood lights of the Boston Airport were a very welcome sight. The trip was made in one hour thirty minutes, and Mr. Webb was able to see his son immediately before the operation.

***

Harold Taylor got caught in some bad weather flying from Westport, NY, to his landing field on the West Shore of Milton. He told us the next day that he held up the clouds with one hand, pushed down on the fog with the other, and flew the plane with his knees. When the airspeed dropped, he knew the wheels were dragging in the water.

***

Harold flew Capt. Smiles O'Timmins over Champlain Valley Fair on three successive days to do his parachute jump before the grandstand. We drove over afterward to collect the money, which took quite a while. My mother had fixed steak for supper and was very upset that we were so late.

***

When Warren R. Austin was running for reelection to the Senate, Harold dropped literature over town. He also flew over the major towns and cities dropping leaflets.

***

Memorial Day, the Fourth of July, and Armistice Day, Harold was asked to fly over Battery Park to help with the celebration. He sometimes had red, white, and blue streamers.

***

Harold was so proud when I got my private license on May 12, 1938. I became the first licensed woman pilot in the state of Vermont. I had my student permit since 1932 but did not get my license for another six years as the physical exam cost too much:

$10.00. That was a lot of money in those days, and you had to have a physical every single year. The pilot's license was issued from the Department of Motor Vehicles. Two other women students had soloed here: Irene Warner on July 22, 1934, and Charlotte Adams on January 23, 1937 on Shelburne Pond.

* * *

My first airplane ride had been in November 1931 with Walton Waite in a Stinson Jr. from the S. Reed Whitney Flying Service of Brattleboro. We flew over a UVM football game. The same plane was bought by Chuck O'Connor of Westfield, CT, and was rebuilt over and over. Before he died, the plane was donated to an airplane museum.

* * *

The summer of 1934 was a time for breaking records: marathon dances, flag pole sitting, flying the farthest non-stop, flying the highest, flying the farthest upside down, staying up in a plane the longest, and refueling in the air.

* * *

Harold was very excited when he got back from a charter to Roosevelt Field in early July 1933. He had been just in time to snap some pictures of Balbo's fleet landing on Long Island Sound. Gen Italo Balbo, Italy's Air Minister, was leading a formation of twenty-four Italian seaplanes from Italy to Chicago, 6,100 miles in seven stops. The fleet also flew back to Italy.

* * *

Every night the fellows would go out with a little cart loaded with construction-type round kerosene bombs to light the runway. Occasionally, a pick-up truck was used. After the airline had passed over, they would go out, snuff the bombs, and pick them up again.

* * *

One time a pilot landed in the meadow of our family farm and the plane landed on its nose. Art Ashley called Harold to ask, "What constitutes a good landing?" Harold replied, "Any one you can walk away from." Art responded, "Well, we just had a beaut' here in the meadow as the plane nosed into the ground but everyone walked away unhurt."

* * *

In 1941, the field lights were electric with above-the-ground wiring. During the day, trucks working on the field would travel over the wires. When night came, the lights wouldn't go on and somebody would have to go out and splice the wires or repair the shredded insulation. Often there were several breaks.

City Light had a small brick transformer house south of the wooden hangar. One rainy night, Harold went out to repair the lights. Meanwhile, the airline dispatcher had reported the breakage to the City Light Office in town. The repairman came out and flipped the switch just as Harold was cutting the wire to splice it. The repairman knew immediately there was a load on the line and cut the switch. Harold's arm was useless for a few days.

* * *

Our tan German Shepherd, Jo-Jo, was as much a part of the airport as anybody. Considered the airport dog, the pilots loved to play with him, and he never tired of it. He would put on a snarling face whenever someone pointed a finger at him and said, "Smile." Pilot Hazen Bean always brought a rubber bone for him to chew, and Jo-Jo seemed to sense which days Beanie was coming in.

Jo-Jo must have been about eleven years old when he got sick and disappeared. Meantime, we had acquired a small Water Spaniel named Brownie. The dogs were very good friends. One day Harold said to Brownie, "Brownie, I wonder what became of Jo-Jo—why don't you go find him." Brownie trotted away down over the south end of the field and came back with Jo-Jo's harness. Brownie took them back to the place where Jo-Jo had died and they buried him there.

* * *

Our student, Art, always wore leather boots and a muskrat hat with the ear-flaps tied up. One spring morning, there was ice on the puddle that often lingered at the intersection. Harold told Art to be sure to land on the other side of the puddle, but Art landed right where he always did—at the intersection. Over went the little yellow plane with its wheels in the air. Art immediately unbuckled his seat belt and landed head-first in the icy water. He stomped toward the hangar swearing like a pirate, flailing his hat against his legs to get rid of the ice and water. George and Loren had the plane flying in a day or two, but Art didn't come back until his hat was dry.

* * *

Mr. F. W. Shepardson and some other businessmen bought the original cornfield that was chosen for the airport. They held it until the city was ready to buy it. The $100 a year rental fee paid by the city was actually the interest on the money they had invested.

* * *

In 1934, Neff owned a Parks P.20. He forgot to check the water in the radiator. When the engine heated up and failed, he had a crash and landed in the Onion River area northeast of the airport. Neff was not hurt, but the poor Parks P.20 never recovered.

* * *

Ken and Eleanor Tomlinson stopped at the airport in the spring of 1932. They were riding a motorcycle and had just come from Swanton where he had taken instruction from Ralph Stancliffe. I find I haven't said much about him but that is typical of Ken—never demanding or aggressive, always capable, dependable, and quiet. He was as much a part of the airport as the sand, the grass, and the many planes he flew. A list of his students would look like the pages of a telephone book. They all remember him as always being calm, patient, and good-natured.

* * *

Cora Rock, wife of the Bristol airport manager, tells a cute story that's a shining example of the way things were back then but can never be again. One day, her husband Joe needed her at the airport and tried to get her by phone. She was in their home's backyard and didn't hear the phone ring. He hopped into his airplane, and as he flew low over the house, he throttled back and yelled, "Come to the airport." She got in the car and went right over, as did just about everybody else in Bristol. That afternoon turned out to be the busiest they had had in two months.

* * *

We purchased an Aeronca taildragger airplane. Harold flew us over to Swanton Airport to pick up the plane to bring to the Burlington Airport. I climbed into the pilot seat of the newly ac-quired airplane. I had trouble taking off toward the hangar as the tail was so heavy, barely skipping over the hangar. Once in the air, I had trouble gaining altitude. Harold, flying next to me, kept mo-tioning me to push the throttle. I motioned that I had. Then he motioned toward me to trim. I motioned that I had. Finally, ar-riving at the Burlington Airport, I was supposed to use the east-west runway, but because of the situation, I decided to go straight in. The crew fixing the north-west runway with rollers, trucks, and other equipment saw me coming and all ran for the woods. I landed safely. Harold immediately ordered the airplane to be checked. They discovered that the tail skid had loosened the fabric around the skid, allowing the accumulation of soil. They vacuumed out a peck of dirt.

* * *

Norm Grady recalls that when he was in high school in 1928, Fred (Freddie) Borgue was flying a WACO 9 (Ox5) biplane made for carrying two passengers in the front and was not outfitted for dual instruction. Flying from the old Middlebury Airport along Route 2 south of town, Norm wanted to fly, and since he had made himself so useful to Freddie, they tried to do something to make it work. People were doing a lot of ice fishing, and in a tackle box, Norm and Freddie found a shiny metal spinner about eight inches long. They put a bracket on the side of the fuselage and bolted it on. Then they drilled the spinner in the center and attached that to the bracket. They got a wooden knob from a kettle cover and fastened that on one end that stuck out toward the pilot. At the other end was the choke control (Bowden wire control), and they fastened that back to the rear throttle.

The rudders were a problem. The rudders in the back cockpit were already the wooden bar type, so they found an old sled and took the cross-steering piece from that and fastened it to the floor with another bracket. From that, they ran a piece of rope to the sled rudder bar.

Now the front cockpit had to have a control stick, and they had quite a problem there. They got a baseball bat and turned down the business end of it to fit the universal joint. Then they ran a walking beam from the front stick back to the rear. (A walking beam is a bar or piece of tubing connecting the two.) The stick was for the elevators, up and down, and gave lateral control on the ailerons, which helped with the banking in turns.

They never did find a way to close the radiator shutters from the front cockpit, and there was no temperature gauge there either. There were no other instruments or gauges in the front, and the instructor had to depend on the student in the back cockpit. The shutters in the radiator had to be closed to get a temperature of 180 degrees. One time he forgot, and the pilot got hot water down his back.

This went on for one whole summer. Fred was also carrying passengers, so the makeshift dual control had to be taken in and out all the time.

It was a full-time job just keeping the engine running. If it turned 1350 rpm or 1300 rpm on the ground, it was considered okay.

One time the throttle wire broke. Another time the radiator shutter control broke and they had to land in a field. They went to the farmers' rain barrel with a bucket and filled the radiator again after the control wire had been repaired.

* * *

On May 19, 1938, Harold helped celebrate Air Mail Week by picking up the very first airmail flying to Swanton, Plattsburgh, NY and Westport, NY, delivering it to the regular airmail plane going from Burlington to Boston.

BIOGRAPHIES

HAROLD W. PUGH

Harold W. Pugh was born on August 11, 1900, in Oxford, Pennsylvania.

Harold's youth was spent running a farm while his dad was a cashier at a bank. He never thought about flying in those days. He evolved, instead, an idea for raising fur-bearing animals; wrote articles about it for *Field & Stream* and other outdoor magazines; and became one of the country's first to operate a fur game farm.

The US Government liked his ideas and results so well that they made him manager of the Government Experimental Fur Game Farm when he was eighteen. He raised 4,000 fox, mink, skunks, and racoons for their pelts to line flying suits during World War I. Pugh wanted to join the war effort as a soldier, but they insisted he was of more value there than anywhere else.

Then he drifted into Westinghouse running a lathe. Then he bought and sold dogs for a Philadelphia pet shop. He was signal painting foreman for the Pennsylvania Railroad for a time. Then he started his own interior and exterior decorating shop in Keeseville, NY. Then the Depression caught him in 1929. With no other revenue source available, he repainted cars for bootleggers, often changing car colors from black to red or green. This is how he earned money to take flying lessons at the Mobodo Airport in Plattsburgh, NY. He received his limited commercial license in 1931 and his transport license in 1932, both being awarded in Burlington. After receiving his pilots' license, he immediately took up the task of teaching others to fly through the Ausable Forks Flying Club at the Plattsburgh airfield. He became manager of the Burlington Airport in 1932. The next year, he got a leave of absence to be the temporary instructor of aeronautic engineers and flight training at the University of Alabama.

Harold's acceptance of the management job at Burlington came at a very critical time in the history of the airfield. It was a period of rapid growth. During these years, the airport grew from a not-very-good cornfield to what was considered the second-best airport in New England in 1941.The runways were lengthened and paved, hangars and administration buildings were built, and various safety and modernization improvements were added.

Having made the acquaintance of the previous airport manager, Arthur Ashley, Harold met Grace Hall, Ashley's step-daughter. Harold and Grace were married in 1934. Together they dedicated themselves to the success of the airport over the next ten years.

In 1943, Pugh resigned as Burlington Airport manager. Harold left aviation and pursued other areas, including Pugh Construction Company.

He died on June 13, 1960, at the age of 59. He left behind his wife, Grace, daughters Barbara Pugh Shimeld, Patricia Pugh Soutiere, and Lucille Pugh McCullough.

GRACE HALL PUGH

Grace was born on May 20, 1908 in Weybridge, Vermont. Her father passed away in 1920 when Grace and her two brothers, John and George, were quite young. They moved with their mother to the family farm, Bay View Farm in North Ferrisburg, Vermont. She attended school in Vergennes. They managed the family farm, renting cottages and taking in boarders in their home. Her mother, Alice Kimball Hall, married Arthur Ashley in 1927. Grace decided to become a teacher, which would allow her to have her summers off to help her mother with the farm, renting out cottages on the lake, and taking in boarders. Grace attended the University of Vermont to become certified as an elementary school teacher. In 1927, she began teaching in rural one-room schoolhouses in Ferrisburg, Vergennes, Waterbury, and Colchester. As a teacher in a rural one-room schoolhouse, she would have to board around with the families of her students, often having to sleep in the bed with the kids. In winters, she would have to trudge through the snow to get to the schoolhouse, start a fire, and shovel the walk before the students came. Her salary was $100 per year.

Grace had her first airplane ride in November 1931. For the twenty-three year-old elementary school teacher, her first flight was made under the watchful eyes of Arthur Ashley, manager of the fledgling Burlington Municipal Airport. The flight was safe and uneventful.

Her second flight was with Harold Pugh whom she met through her step-father, Arthur Ashley, when he was invited to the family farm. That flight was her impetus to fly. In 1934, the couple married and together over the next ten years they forged the Burlington Airport. Grace was a ground school instructor for the Civilian Pilot Training Program at the University of Vermont from 1940-1943. She also taught ground school in Vermont at St. Albans, Swanton, Norwich University, Champlain Airport and in Plattsburgh and Keeseville, NY. She taught meteorology for the Ferry Command and assisted with the training of pilots through the Fli-Rite School of Aviation.

Grace became the first woman pilot to be licensed in Vermont on May 12, 1938. She had her own gray Taylor Craft nicknamed the Mouse. It had a 40-horsepower engine, no brakes, no radio, no rear wheel, only two or three instruments, and a stick control. From 1932 to 1945, Grace was totally involved with the Burlington Airport.

After her years at the airport, she raised her two girls, Patricia and Lucille, and returned to teaching in 1949. She taught in Colchester, and in 1953, she went to South Burlington to teach second and third grades. She was recognized with teaching awards over the years, and during her school teaching career, she taught over 2,000 students. The former students declared her their most formative teacher. Grace retired from teaching in 1972.

In 1969, Grace, along with her brother, George Hall, were honored by the U.S. Air Force Thunderbirds with a fly-over. A reception at Marble Island followed.

Thunderbirds, 1969

From 1976 to 1986, she volunteered as a tour guide at the Burlington Airport. She made significant contributions to the book *Burlington International Airport: A Pictorial History*, published in 1982.

In May of 1988, on her 80th birthday, she was celebrated in St. Albans at the Franklin County Airport. The event was sponsored by the Experimental Aircraft Association. Hundreds attended, including past students, aviation friends, and family.

Grace died on November 14, 1996. She left behind her daughters, Patricia Soutiere and Lucille McCullough, stepdaughter Barbara Pugh Shimeld, grandchildren, Lisa McCullough, Cam McCullough Sato, Abbie McCullough Bowker, Natalia Soutiere-Burud, Shawn Soutiere, and great grandchildren.

She never lost her passion for aviation.

Eighty-year-old Grace sill in love with flying

ADDENDUM

"Is Flying Safe Today?"
Chamber of Commerce Speech 1930
By Harold W. Pugh

This is called the age of the air, but it is surprising when you think of it how small a proportion of the American people have yet to come of this age. By many persons, flying is still considered the "Great Adventure," and though the sky is dotted with planes, they keep their feet firmly fixed on the ground. Yet they will cheerfully submit to the rush of being "done in" by automobiles—and it is a risk as great as many soldiers took in the trenches—but "Go Up?" No Siree! TOO DANGEROUS!

The truth is, of course, that despite a few sensational crashes—and all the crashes seem to be reported sensationally—flying is safe and steadily growing safer. *The London Daily Express* says that only one fatal accident occurs on the British Commercial Airlines in 2,300,036 miles flown, and we do much better than the British.

Americans record is 38,000,000 miles flown at the cost of a life. Now with this record, I'll leave it to you. Is flying dangerous? Well, when planes leave the airports with the regularity of railroad trains and arrive at their destinations just as regularly, day and night, it is a bit difficult to see just where the danger is. But old ideas are hard to shake. There are even old timers who have never ventured on the "steam cars" and who wouldn't think of getting aboard an automobile just because their grandfathers wouldn't.

Fifty years from now, there will be some persons who will keep off the planes for the same reason and perhaps slip on the stairs and break their necks. But—"That's different!"

The lives of people always have been founded by the radius of their transportation. At one time, the distance a horse could travel in a day was the range upon which men predicted their activities. Railroads and motor cars shrank distances and crowded years into months. Now airplanes make it possible to go even farther in less time. Social contacts and business calculations are now expanding to the wider radius of flying. It is not exaggerating to say that airplanes will remake the map of America, introducing even greater changes than the transition from ox cart to railroads.

While airplanes make possible totally better living and business conditions, they are mechanical vehicles of transportation and as such are not immune to restriction. An ocean liner, a locomotive, or a motor car are as safe as the person in command of it. So is an airplane. Ocean liners must watch storm warnings and often lie anchored until fog clears. A railroad engineer must continually obey signals. No motor car owner can drive safely if he ignores traffic regulations and road conditions. We believe that flying is actually safer when it is regulated by the same considerations and common sense required in other means of transportation. The sooner this 'common sense' view of flying is accepted and the "romance of being bird men" is forgotten, the sooner the general public will enjoy the utility benefits possible through flying. The sturdy, safe, and dependable aircraft of today are capable of day in and day out service, proven by the regularity with which they meet transportation schedules for all kinds of purposes in all parts of the world. They afford the same comfort and luxuries to which people are accustomed to in land conveyances.

Flying is no longer a prophecy, it is a reality today. Although some people will never enjoy this means of travel, aviation is here and here to stay.

Being an airport manager, a pilot, and a dispatcher for Boston-Maine and Central Vermont Airways, I will say a few words concerning the traffic end of this form of transportation. The men in the traffic department of an airline whose business it is to educate the traveling public must be able to create and maintain favorable impressions and build up the faith and confidence in the mind of their prospective passengers. The routine of handling passengers and supplying them with tickets, dispatching express airmail, answering questions, checking weather reports, working up daily financial accounts is in itself but a part of the efficient methods which go towards keeping an airline in the air. Behind the routine in the traffic and publicity departments lies a still more important part of the great work getting people into the air who are still groundlings.

Last year, one airline company alone handled 568,940 people. While this number shows a steady increase in air passenger traffic, it is but a small percentage of the number of people who travel and could use the airlines to their advantage. It is a constant struggle to increase passengers and express revenue, the eternal battle against prejudice, ignorance, and travel habits of long standing, which adds the final touch of spice and zest to the adventure of air transportation.

We must maintain a knowledge of psychology and a sixth sense which indicates the correct time and line of approach to a prospective air traveler. Where other means have failed or only half succeeded, a sudden turn of events and the swift grasping of the opportunity offered has a thousand times delivered a new convert into the air.

There was one instance in which a man just could not see the sense of air travel. He had been up once with a joy ride pilot in an antiquated plane, and it ended up in a corn field with the plane on its nose. The man was quoted, "There were too many risks anyway and why did a man have to be in such a hurry." He did agree that all that could be said of air travel was probably true and that there was no comparison between the ship he had been up in years before and the modern airplanes, but he was not interested. The case was obviously one of fear. The only way to cure him of it was to get him up in a modern plane and show him first-hand what air transportation of today has to offer. He reluctantly agreed to take a flight over the city in one the newest and finest airplanes built today.

It is a long step, however, from a man's acceptance of a short flight and the moment when he is comfortably seated in a spacious cabin and several thousand feet in the air and wondering just what in heavens name he had been afraid of. More than one has failed to show up for the flight or else had become too busy to get away from the office for it.

It was learned that the man's wife had been teasing him to let her fly to New York. There are a few men with whom wives do not usually get their way. It so happened that she got her say after all. The details and excellent qualities of the planes were brought to his attention, and once more the working of an airways system was explained. He became more interested in flight, he asked questions, and at last he decided he would let her fly one-way. It was then brought to his attention that there was a ten percent reduction on a round-trip ticket, and if she liked the journey, she would undoubtedly want to fly back. If not, the difference between the one-way fare and the round-trip fare would be refunded. He bought a round-trip ticket. When his wife arrived in New York, she sent him a telegram. "I will never travel any other way again, returning by air."

The telegram was followed by a longer descriptive letter of the flight. Today the man flies whenever he travels and enjoys it thoroughly. And so will anyone of you once you try it.

Thank you, and I hope I have impressed you with the safety of air travel and that you, too, will all fly soon.

FLY
by Harold W. Pugh

I dare you to fly above the clouds so silky below, with mountains protruding from below and tell me you weren't touched.

I dare you to see the fog laying in the mountain valley just at sun rise, the scattering of pinks and reds and tell me you aren't moved.

I dare you to slide through the air by yourself and look down on it all, the feeling of serenity and peace and insignificance and tell me you didn't feel something special.

I dare you to fly with the birds and break your earthly bonds and tell me you didn't enjoy, perhaps for the first time, something special, unique, and so personal.

A very special place this sky, so many moods and so many feelings. Life is too short not to enjoy, and I dare you to fly.

FLYING DAYS ARE DONE
by Harold Pugh

My flying days are over
My helmet laid away
My wings are clipped close to my sides
My dreams have gone astray
No longer can I, like the gull
Soar, dive, and fly
No longer can I chase the clouds
Chained to the earth am I.
How well my mind still visions
My classmates eager, true
They toss their hearts up to the sky
To chase it in the blue
Soaring, climbing, banking, diving.
Graceful birdlike things
Oh, how I envy those who fly
I wish that I had my wings.
Yet I know 'tis not for me
My niche I haven't found
Perhaps 'tis written in the book
That I stay on the ground.

ACKNOWLEDGEMENTS

Thanks go to Grace Pugh for her foresight and diligence in documenting this early history of the Burlington Airport as she witnessed it. Through many years of dedication and hard work alongside her husband, Harold Pugh, they created their legacy. The several photo albums and scrapbooks she maintained are now is safe keeping at the University of Vermont Library, the Silver Special Collections for public access.

A special thank you to my sister, Patricia Soutiere, for her help and participation in making this book a reality.

Thanks to the Chittenden County Historical Society for their generous grant to assist in the publication of this book along with their wonderful show of support.

Thank you, Cam Sato (Grace and Harold's Granddaughter), for employing your keen eye for details: proofreading, editing, and making insightful recommendations.

Thank you, Newt Bowker (Grace and Harold's Great-grandson), for the creation and design of the book cover, as well as employing your photography skills to improve and digitize the photos.

Thank you, Onion River Press staff, for your patience and expertise helping us through the publication process.

Thank you, friends and family, for providing photos, stories, information, proofing, and support of this publication.

www.ingramcontent.com/pod-product-compliance
Lightning Source LLC
Chambersburg PA
CBHW021335060726
47591CB00006B/2028